Monographs in Electrical and Electronic Engineering

Editors: P. Hammond and D. Walsh

Energy Methods in Electromagnetism

P. HAMMOND

CLARENDON PRESS · OXFORD
1981

Oxford University Press, Walton Street, Oxford OX2 6DP

OXFORD LONDON GLASGOW
NEW YORK TORONTO MELBOURNE WELLINGTON
KUALA LUMPUR SINGAPORE JAKARTA HONG KONG TOKYO
DELHI BOMBAY CALCUTTA MADRAS KARACHI
NAIROBI DAR ES SALAAM CAPE TOWN

© P. Hammond 1981

Published in the United States
by Oxford University Press, New York

British Library Cataloguing in Publication Data
Hammond, P
  Energy methods in electromagnetism.
  —(Monographs in electrical and electronic
  engineering).
  1. Electromagnetism
  I. Title   II. Series
  537   QC760   80-41146
  ISBN   0-19-859328-7

Filmset in Great Britain in 'Monophoto' Times New Roman
by Eta Services (Typesetters) Ltd., Beccles, Suffolk

Printed in Great Britain, by
Lowe and Brydone Printers Ltd., Thetford, Norfolk

'We avail ourselves of the labours of the mathematicians, and retranslate their results from the language of the calculus into the language of dynamics, so that our words may call up the mental image, not of some algebraical process, but of some property of moving bodies.'

James Clerk Maxwell: *A Treatise on Electricity and Magnetism*. Article 554.

Preface

Electrical devices and systems are generally described either in terms of circuit parameters or in terms of fields distributed in space. This book deals with the relationship between fields and circuits, a subject which seems to have received relatively little attention and which has puzzled me greatly for many years. When I began to think about it, I found that energy could be used as a unifying principle, but I had to study variational mechanics before I saw that the concept I was looking for was 'energy at equilibrium'. Although my exploration of this subject is far from complete I hope that this book will be found helpful by the reader who is himself exploring these matters.

To some people this talk of exploration will sound strange. I have heard it said that electromagnetism is completely closed as a subject, because Maxwell's equations define all that can be known about it. Although I have the utmost respect for these equations and for the man who discovered them, I am sure that electromagnetism is far from closed. Maxwell's equations are an invitation to travel abroad, just as maps are, and there is a good deal of difference between looking at a map and travelling in foreign countries. Even a cursory reading of Maxwell's *Electricity and Magnetism* will show that this work is more like an explorer's diary than an encyclopaedia.

Some readers will feel that, while this may be so, the proof of the pudding is in the eating and that the improved understanding which they may gain from this book must be matched by improved methods of calculation and design. I agree with that sentiment and hope that the book will be useful in this respect also. There are a considerable number of worked examples, which should be sufficient to enable the interested reader to apply the energy method to problems within his own speciality.

My thanks are due to many people, including all those who, when they came to see me on some other matter, found themselves subjected to a long discourse on electromagnetism and bore it cheerfully. By name I can mention only a few: the late Professor E. B. Moullin, who first shared his enthusiasm for electromagnetic problems with me some thirty years ago, Dr. G. K. Cambrell who introduced me to the algebra of vector spaces, Dr. F. J.

Evans, a fellow-explorer who first told me of Lanczos' *Variational Principles of Mechanics*, Professor M. J. Sewell, who sent me his valuable papers on convex and concave functionals, Dr. J. Penman who helped in the preparation of two papers published in the *Proceedings* of the I.E.E., Professor G. Rodriguez-Izquierdo who criticised incisively the methods proposed in those papers, Drs. J. B. Davies, D. R. Farrier, and P. J. Tavner who read the typescript and made many valuable comments, and Dr. R. L. Stoll, friend and fellow-worker for many years. I also want to thank Miss S. D. Makin for her patient and expert typing and my wife for suffering electromagnetism to invade our house, while this book was being written.

P.H.

Southampton
Easter 1980

Contents

List of principal symbols

A	action
A	magnetic vector potential
B	magnetic flux density
C	capacitance
C	vector function
D	electric flux density
E	electric field strength
E	electromotive force
F	force
G	content, Green's function
H	magnetic field strength
I	electric current
J	electric current density
J	moment of inertia
L	Lagrangian energy, inductance, operator
M	mutual inductance
P	power
P	vector function
Q	vector function
Q	electric charge
Q^*	magnetic pole strength
R	resistance, reactive power
T	kinetic energy, torque
U	potential energy
V	electric potential, potential difference
W	energy
X	reactance
Y	convex energy functional
Z	concave energy functional

a   acceleration

b damping constant

f force, basis function

g conductance, source distribution

k spring constant

m mass

p direct stress, general momentum

q general coordinate

$\dot{q}$ general velocity

r resistance

t time

v   velocity

w work function

x reactance

γ propagation constant

ε_0 permittivity of free space

ε_r relative permittivity

ε $\varepsilon_0\varepsilon_r$

$\lambda, \boldsymbol{\lambda}$ Lagrange's multiplier

μ_0 permeability of free space

μ_r relative permeability

μ $\mu_0\mu_r$

ρ volume charge density

ρ_m volume pole density

σ surface charge density

τ shear stress

ϕ electric scalar potential

ϕ_m magnetic scalar potential

χ scalar potential

ψ scalar potential

ω angular frequency

Δ eddy current skin depth

Φ magnetic flux

Ψ electric flux

1. Electromagnetism and Mechanics

1.1. Introduction

The purpose of this book is to show that a method of looking at electromagnetic phenomena from the standpoint of energy has much to recommend it. The three advantages which arise from this approach are first that the conceptual framework of the subject is simplified, second that particularly simple methods of calculation can be devised from energy considerations, and third that these methods of calculation are directly related to measurement techniques. In the book we shall concentrate chiefly on the first and second advantages, namely on concepts and calculations. Since the subject of electromagnetism is vast in extent and since the author's experience is limited, the book is intended as an introduction to a point of view rather than as a handbook of useful results. Nevertheless it is hoped that the reader will find sufficient help to enable him to apply the method to his own field of interest.

We must of course admit straight away that there is nothing radically new in looking at electromagnetism from an energy point of view. All books on the subject include energy theorems such as Thomson's theorem, which we shall discuss in detail in Chapter 4. However, what is perhaps different about this book is that we regard these theorems as central, whereas energy methods are generally treated as being somewhat peripheral to the main development. An example will illustrate what we have in mind. Suppose we think of electrostatic problems, which are often considered to require the calculation of the field at every point of the region of interest. In general this requires a subdivision of such a region into a fine mesh of field parameters which have to be calculated by means of a numerical scheme. The calculation is complete when all the local values have converged to the actual values. Central to such a scheme is the concept of a potential function at a point. It is our contention that this may be a wasteful method because it generates information which is unlikely to be of use either to the manufacturer or the operator of a device. It is very seldom that anybody requires to know a potential distribution. Such a distribution would have to be measured with a very fine probe and even then a point function is unobtainable. Generally what is required is a knowledge of the energy distribution, expressed perhaps in terms of capacitance, and a fairly coarse distribution will be sufficient. The energy method of this book deliberately restricts the information to that which is required. Instead of focusing attention on a point potential it works with the energy of the whole device treated as a system. By noting that the system energy at equilibrium has a maximum or minimum value we can obtain relatively simple approximations to the energy distribution. These approximations provide both upper and lower bounds to the accurate values and so establish confidence limits which can be set by the engineer in accordance with his needs in any particular application.

What has been said here about an electrostatic problem has far wider applications. It applies equally well to magnetostatics and to problems concerned with electromagnetic induction and radiation. Equally well it applies to network problems, where the energy distribution is often more important than a knowledge of all the individual currents and voltages. From a measurement point of view it is of course apparent that the energy level is the critical factor.

We hope that these practical considerations of calculation and measurement will encourage the reader to study the method. It hardly needs saying that energy methods are unlikely to displace other methods of calculation. All that they will do is to increase the choice of method and to reduce the complexity of a calculation in appropriate cases. This means that the energy method can be applied to problems of complicated geometry and to problems containing non-linear parameters which present particular difficulties in field calculations. It also means that less complex problems can be transferred from a large digital computer to a desk machine with a corresponding reduction in time and cost. This might be especially useful in the early stages of a design process and could be followed by a more detailed and accurate computation when the design is nearing completion.

In this introduction it is impossible to develop a convincing argument that the energy approach to electromagnetism simplifies the conceptual structure of the subject. This will become apparent in subsequent chapters. To the author, who is primarily a teacher, the conceptual simplicity of the approach is of overriding importance. In popular usage theory is often contrasted with practice, but in science and engineering theory and practice are closely joined. The use of design formulae without theoretical backing is unsatisfactory and dangerous. At best it can be justified where the design of a device is well established, but when changes have to be made it is necessary to go back to first principles. It is then of the greatest importance that these principles should be as simple as possible, not only because this will enable an individual designer to think out improvements, but also in order to provide a simple language in which ideas can be interchanged.

For this reason much of this book is concerned to provide the reader with an understanding of the physical principles of electromagnetism which underlie the method of calculation. The principles are few in number and can be understood without much difficulty. A grasp of the principles will enable us to compare different processes and to obtain quickly a qualitative understanding of particular devices. Such understanding results in a great saving of mental effort and enables new developments to be incorporated into one's mental framework of ideas. This provides mental stability, which is a valuable possession especially in dealing with a rapidly developing subject.

The needs and interests of different readers will vary. It is a convention that a book should be read from beginning to end and in a detective story one should resist the temptation to turn to the last chapter to find the solution. However, in a book like this there is no need to proceed in any particular manner. Learning is generally a cyclical process. To understand a matter we need to look at it many times and from different angles. The reader who is interested chiefly in application may find it best to turn at once to Chapters 4–6 and later to Chapters 2 and 3, but the reader who is interested chiefly in the structure of the subject may find it preferable to follow the order of the book. The level of treatment is such that it should be suitable for students in the final year of a degree course or during postgraduate studies, but the book is intended also for practising engineers, especially for those concerned with the design and analysis of electromagnetic devices.

1.2. Historical background

We have already mentioned that this book does not need to be read in strict sequence, and this is particularly true with reference to this section which can easily be omitted on a first reading. The section will interest only those readers who, like the author, are fascinated by the history of ideas. How did the notions of electric charge and field arise and how is electromagnetism related to other branches of physical science or to science in general? Answers to these questions should give insight into the use of the various concepts. We have inherited a toolkit of ideas and may be helped in the selection of appropriate tools if we consult the people who developed them in the first instance. Although the ideas underlying the subject are few in number, so that the language of electromagnetism employs a relatively small number of words, this certainly does not mean that the construction of this language was a simple matter. Indeed the opposite is true and we have inherited something which is the result of the labours of men of genius over a period of many centuries. A historical approach, although it is somewhat out of fashion, will help us to gain a perspective and will also anchor new developments in past experience. It is of course true that the scaffolding used in construction should be removed to show the beauty of the completed building, but the analogy is misleading. In the development of physical concepts the scaffolding and the building are the same thing. Insight into the meaning of concepts can best be acquired by looking at the way in which these concepts were selected by the great pioneers of the subject.

Although the Babylonians and Egyptians made progress in technology and the practical application of mechanics, they did not formulate general principles. It was the genius of the Greeks to be particularly concerned with giving a coherent account of the phenomena around them and most of our present scientific ideas can be traced to Greek thought.

The first great name in Greek science is that of Thales of Miletus (around 600 B.C.). To him is attributed the discovery of electrostatic attraction. The word electricity is derived from the Greek word for amber, a substance which was used in these early experiments. It is interesting that a subject which was to be developed only in modern times should have its roots right at the beginning of history. More important than Thales' observation of electrostatic effects was his contribution to scientific thought. He postulated that the multitude of physical phenomena must be related to a single invariant entity. With this bold conjecture he charted the course which science has taken ever since, namely the search for invariants, or conservation laws, such as the conservation of matter or of energy. By this means scientists have been able to reduce the number of hypotheses and devise laws of ever-increasing generality. Thales chose water as his universal substance or principle. His contemporaries Anaximenes and Anaximander accepted the idea but were less specific. The former chose air, and the Greek word could also mean breath or spirit. The latter deliberately avoided the name for his substance which was the substratum of all that could be observed. He might well have been satisfied with our present idea of a space–time continuum.

Having defined an invariant substance these philosophers explained observed change in terms of the motion of substance. Motion was taken as a basic idea not needing further explanation. Ever since the time of these three philosophers of Miletus matter and motion have been the foundation stones of physical science.

The idea of force as being the cause of motion was developed a century latter by Empedocles. He defined both attractive and repulsive forces and described centrifugal force. These ideas fit well into the modern view of a universal gravitational attraction and cosmic repulsion. Attractive and repulsive forces are also dominant in electrostatics and magnetostatics. Another of Empedocles' important contributions was the 'exclusion principle' that two bodies cannot be in the same place.

Empedocles also taught that light is propagated through space with a finite velocity. He generalized the idea into a theory that all bodies give off some of their substance as emanations, and he used this as an explanation of the attractive force of a magnet on iron. This is an idea we find again in William Gilbert's book on magnetism published in 1600 A.D.

The Greek thinker who had the most lasting and dominant influence on scientific thought was Aristotle who lived in the fourth century B.C. His scientific views were accepted very widely until the time of the Renaissance and the start of modern science. At that time the settled Aristotelian world-view was a great hindrance to progress. Aristotle, like the other Greek philosophers, was strong in the realm of ideas but did not in general submit the ideas to the test of experiment. The Greeks made little progress in technology and so their science lacked the stimulus of accurate observation. However, they were pre-eminent in conceptual thinking and it would be foolish to dismiss Aristotle's contributions in this realm.

The chief difference between Aristotle's views and our own arises from Aristotle's preoccupation with biological rather than physical ideas. The modern view of physics, which has been so remarkably successful, is a mechanistic view. We confine ourselves to the question 'how' the universe works and deliberately refuse to introduce the questions of purpose prefixed by 'why'. The idea of personality is kept at arm's length and all is impersonal matter and motion. Even the study of living organisms is dominated by this approach. This does not commit us to the view that personality is a fiction, but it means that it is not a term of physical science. Nor is it a term of any other branch of science, if these branches are thought of as subsystems of physics, or at least of being subject to the methodology of physics.

Aristotle, on the other hand, makes purpose the dominant concept in his scheme of things. The regularity of nature which we ascribe to mechanical laws he sees as the purposeful action of mind. This preoccupation superposes an additional layer of concepts on physics and complicates science. Aristotle attempted to solve both physical and metaphysical questions with one scheme and in this optimistic attempt he was bound to fail. Nevertheless it is very worthwhile to take a look at his ideas.

Aristotle's theory of motion is of particular interest. He distinguishes between 'natural' and 'compulsory' motion. The former is either circular or linear. The heavenly bodies move in circular motion because this motion is simple and symmetrical. Circular motion does not need to vary and this makes it pre-eminent. There is also natural linear motion: light bodies move upward and heavy bodies downward. They do not need a force to move them, because force has been replaced by the concept of weight. Aristotle concluded wrongly that the velocity of a falling body is proportional to its weight.

All other motion is compulsory and needs force, the distance traversed by a body being proportional to the force and inversely proportional to the mass of the body. About 2000 years elapsed before these laws of motion were put on to a satisfactory basis by Galileo and Newton. This testifies to the difficulty of the problem. The ideas of natural and compulsory motion have been abandoned, but it is interesting to observe that they reflect the two aspects of mass, namely its gravitational and inertial properties. It is only in Einstein's general relativity theory that these two separate properties have been unified.

One other interesting idea of Aristotle occurs in his discussion of the impossibility of a vacuum. He does not allow the distinction between geometry and matter. If an object were to be surrounded by vacuum it would in accordance with his ideas have no interaction with anything else and it would be impossible to ascribe to it a position or place with respect to the rest of the world. This is a denial of the possibility of 'action at a distance' stated in the strongest possible form. It makes geometry into a feature of matter instead of regarding it as an abstract framework which determines the relative position of material objects. This theory represents a remarkable foreshadowing of the modern view in which the metric of space–time is dependent on the presence of matter. However, even before the formulation of general relativity theory Aristotle's view of action through contact had a profound influence. It was the guiding principle of the world picture of Descartes who postulated that space was filled with a substance which, although it could not be observed directly, was capable of transmitting force and energy. This substance was given the Greek name aether, a word meaning the upper air. Aether theories have been prominent in the development of the scientific description of electricity and magnetism, although the word itself has dropped out of use. The present-day ideas of electromagnetic fields are closely related to this concept.

Aristotle's views were challenged by the Greek atomist school, whose teaching is associated with the name of Epicurus, a younger contemporary of Aristotle. The atomist universe consisted of a vacuum filled with solid particles. These particles were of the same substance but had different shapes. All these atoms were thought to be in a state of permanent violent movement. Force was transmitted by impact and action at a distance was explicitly denied. Even vision was due to the impact on our eyes of material emanations from the objects which the eyes see. The relationship between Greek and modern atomic theory is fascinating but outside our present interest of tracing the development of electromagnetic and mechanical ideas.

After the brilliance of the Greek period there came centuries during which scientific thought was virtually at a standstill. There are almost two thousand years between Aristotle and the beginning of the modern period of rapid development. The great puzzle of motion was solved by the combined efforts of Galileo and Newton. Where Aristotle had postulated that force was proportional to velocity, a view which could be supported by observation of a horse and cart, Galileo postulated that force was proportional to change of velocity. It is easy to underestimate the revolutionary nature of this difference. It involved the idea of motion without force or contact, motion in an empty Euclidean space stretching to infinity. The idea of inertia, which now seems obvious, would have been to a follower of Aristotle a tangle of absurd notions.

Newton's gravitational theory (1687) produced another great change of outlook. The idea of a universal force of gravitational attraction depending only on the distance between material objects effectively separated geometry from matter. Newton's theory is the first example of a theory of action at a distance. The great advantage of such theories is that they enable attention to be focused on individual objects, whereas the continuum theories look at the whole universe as a single entity. Newton himself had misgivings about action at a distance as being the complete story. He even went so far as to call it an absurdity and suggested that his theory did not touch on the mechanism of gravitation but only gave a mathematical description of the phenomenon. However, in spite of this bow in Aristotle's direction Newton's theory marked a new point of departure. In particular his method was radically different from that of Descartes which was dominant in France. Voltaire wrote humorously in 1730: 'A Frenchman who arrives in London will find Philosophy, like everything else, very much changed there. He had left the world a plenum, and now finds it a vacuum.'

The history of electricity and magnetism followed a parallel development. William Gilbert's important book *De Magnete* published in 1600 ascribes electrostatic forces to emanations from electrified bodies. These emanations were material substances but of a very tenuous nature. Gilbert compared them to a scent which could be given out by a body for a long time without causing an appreciable loss of weight. He was less definite about magnetic forces because these were much stronger than electric ones. He ascribed them to the fact that magnets were surrounded by a region of magnetic strength, as we should say by a magnetic field. Gilbert explored the field with a small magnetic needle and noted the direction as well as the strength of the vector field of force. He also drew attention to the importance of the poles of his magnets and postulated the idea that the earth was itself a great dipole.

Gilbert's theory was a continuum theory. Like all such field theories it was difficult to quantify and lacked the simplicity of Newton's gravitational formula. It was not until electricity and magnetism were brought into the Newtonian framework in the middle of the 18th century, about 150 years after Gilbert, that progress became rapid. The way in which this came about is interesting.

Experiments with friction machines suggested that there were two classes of substance, 'electrics' and 'non-electrics'. The former could be electrified but not the latter. The distinction corresponds roughly to that between insulators and conductors. This suggested that the seat of the action lay in the material of the substance itself, a view which was reinforced by the observation that electric effects could be conveyed from one body to another. Thus the idea of an electric fluid in a body began to replace the idea of emanations. It was an idea which was brilliantly, and dangerously, vindicated when Franklin used a kite to draw some of the electric fluid from a thundercloud. Priestley (1766) observed that inside an electrified cup there was little electric field and suggested that this could be explained if the electric fluid had a law of force between its droplets which was the same as Newton's law of gravitation. By means of a famous experiment Cavendish confirmed Priestley's hypothesis. Investigations on magnetism showed that magnetic effects also were closely associated with magnetic objects, as had been shown by Gilbert. Michell (1750) showed that although magnetic poles always occur in pairs the process of calculation of magnetic force is much simplified if each pole is considered separately. If this is done it is found that the law of force between poles is again the Newtonian law of the inverse square of the distance.

After these discoveries gravitation, electricity, and magnetism could all be treated by means of the same mathematical theory based on a principle of action at a distance. The idea of emanations was abandoned and field theories became less prominent.

However, the next wave of discovery reversed all this. Although it was highly satisfactory that one law of interaction had been found for gravitational mass, electric charge and magnetic poles, scientists continued their search for links between these phenomena. Such a search for unity has been a guiding principle from the earliest Greek times, as we have mentioned. The first link between electricity and magnetism was discovered by Oersted (1820) who observed magnetic effects around a wire carrying an electric current. Oersted's observations suggested that the action was in the field rather than a type of action of a distance. In a rapid succession of brilliant investigations Ampère took up Oersted's results and deduced that a small current loop and a magnetic dipole had identical effects. He invented the idea of a current element as a small source of magnetism and developed a law of force between such elements. If the force was assumed to act along the line joining the elements, as in the Newtonian law, Ampère's law had one term varying as the inverse square of the distance and another as the inverse fourth power of the distance. Ampere realized that other forms of his law were possible because experiments necessarily involved complete circuits rather than current elements. An inverse square law could be postulated, but the direction of the force would then depend on the directions of the current elements. It was therefore difficult to incorporate the magnetic action of electric current in a Newtonian scheme. Faraday (1831) discovered the complementary effect to Oersted's experiment. Oersted had found a magnetic effect of electric current and Faraday in his law defined the electric effect of a magnet. He concentrated his attention on the space around the magnets and circuits which he used. He had little mathematical knowledge and the mathematical elegance of Ampère's formulae did not appeal to him. His great gift lay in his unsurpassed physical imagination. He 'saw' the field as a physical entity no less real than the coils of wire with which he was experimenting. To him we owe the quantitative description of the field in terms of tubes of flux. His successor Maxwell clothed Faraday's description in the terms of the vector calculus by which means he was able to construct a mathematical model of the aether which carried the electromagnetic interactions. By this aether model he was able to predict the phenomenon of electromagnetic radiation and to show that light was an electromagnetic phenomenon of a particular frequency range. After Faraday and Maxwell the world was a plenum again rather than a vacuum.

However, after this great advance opinions again began to change. Maxwell himself had realized that an alternative to his aether theory was a theory based on action at a distance, provided that a time delay was introduced into the calculations. The finite speed of electromagnetic propagation had to be inserted into all time-varying effects. For static fields the simple Newtonian formulae remained. Maxwell's successors endeavoured to discover further properties of the aether. In particular it was important to discover its motion. If it was at rest then the earth would have a relative motion to it. The aether would thus provide a frame of reference for the universe. Alternatively the aether might be carried along by the earth. All experiments designed to determine this matter gave no help; the motion of the aether was unobservable. When this conclusion was accepted it produced the revolutionary consequences of Einstein's relativity theory. Time and space were seen to be inextricably mixed not only in electromagnetism, but in ordinary mechanics and indeed in all physical phenomena. It is interesting to notice the close relation between mechanics and electromagnetism in all this development. Priestley's inverse square law of charge sprang from his consideration of a theorem of Newton about the absence of gravitational attraction inside a shell of gravitating matter. Modern relativistic mechanics arises from Maxwell's equations of electromagnetism and the dominant role of the velocity of light in his theory. The concepts of force and energy apply equally in mechanics and electricity, and the link between the subjects is further strengthened by the equivalence between mass and energy.

With the advent of relativity theory the aether has dropped out of use, except as the medium which carries electromagnetic radiation. It might be said that the world is a vacuum again in which there is a distribution of energy, concentrated where there is matter and diffuse elsewhere. However, the interaction of this energy affects the geometry, so that the space is no longer Euclidean. Thus the plenum and vacuum, or field theory and action at a distance theory, have converged.

For the purpose of the argument to be developed in this book it is of great importance to realize that both mechanics and electromagnetism deal with systems and not with isolated objects. It does not matter whether we seek to locate the action in the field or in the sources of the field, but it is essential to define either the extent and boundaries of the field or the totality of the sources before anything can be said about force or energy.

In ordinary mechanics it is often useful to focus attention on a single mass or on a rigid body. The other parts of the system are then replaced by suitable forces. Newton's laws of motion are designed for that very purpose. However, in electrical problems this is seldom the best way because electrified objects are themselves systems with an internal interaction. For this reason we shall use the mechanical method associated with the name of Lagrange which was developed for the study of systems and we shall find that the adaptation of this method to electrical problems leads to a very useful method of calculation.

To us both the field and its sources are parts of a system. Our historical study has shown that there is no answer to the question of whether field or sources are primary. This study has also shown the close relationship between the development of mechanics and electricity. This will give us confidence in applying the methods of mechanics. Better still it will unify our ideas and reduce the number of hypotheses, which is always desirable and particularly so for engineers who have to deal with a wide variety of devices and processes.

1.3. Electromagnetic concepts

In order to highlight the particular point of view of this book we shall find it useful to present a brief review of electromagnetic concepts and relationships. It is convenient to start with the concept of electric charge expressed in the law of force

$$\mathbf{F} = \frac{Q_1 Q_2}{4\pi\varepsilon_0 r^2}\,\hat{\mathbf{r}}. \tag{1.1}$$

The factor $4\pi\varepsilon_0$ arises from the choice of units and dimensions and is in that sense arbitrary. We shall not concern ourselves with it. Our interest is concentrated on the fact that this law of force defines electric charge in terms of an action at a distance between two objects. The force acts along the line joining the two sources. It depends on their magnitude and also on their position. The force can be either repulsive, if the charges are of the same sign, or attractive if the charges are of opposite signs. If the two charges are regarded as a system we note first of all that for equilibrium additional forces need to be applied to oppose the electrostatic forces which by themselves cannot produce equilibrium. This is called Earnshaw's theorem and we shall discuss it more fully in Chapter 4. Secondly we notice that we can associate an energy with the system which is equal to the assembly work of the charges. It is usual to consider two charges of equal sign. Work then has to be done to bring the charges together. The energy is given by

$$U = -\int_{\infty}^{r} \mathbf{F} \cdot d\mathbf{r} = \frac{Q_1 Q_2}{4\pi\varepsilon_0 r}. \tag{1.2}$$

In this statement one of the charges is brought up from infinity. The energy depends on the position of the charges and is therefore a potential energy. It is of course a scalar quantity, and if a system consists of more than two charges the total energy can be obtained by addition. The force on any particular charge can then be obtained by finding the gradient of the energy. This procedure avoids the need for vector addition.

The system defined by equation (1.2) is unbounded because it includes infinity. In general we shall be concerned with bounded systems and in that case it is better to define the assembly work in terms of separating unlike charges rather than bringing together charges of the same sign. We then write

$$U = -\int_{a}^{r} \mathbf{F} \cdot d\mathbf{r} = \frac{Q_1 Q_2}{4\pi\varepsilon_0} \left(\frac{1}{r} - \frac{1}{a}\right). \tag{1.3}$$

The original distance between the charges introduces a constant which does not, however, contribute to the force. It arises from the definition of charge as a point source. If instead we define a charge density per unit volume, we can consider an overlap of positive and negative charge density. If, for example, the densities are equal they would cancel and the assembly work would then produce positive and negative charge from a region of average zero charge. The constant term in the energy expression then disappears. We shall return to this in the discussion of Green's theorem in Chapter 4.

It will have been noted that in the system of like charges we could also exclude infinity. Nevertheless this would not exclude interaction with other charges outside the system. An isolated system must contain no net charge. It is interesting that the total net charge of the universe appears to be zero. In engineering, devices such as capacitors also generally have zero net charge. This enables us to treat them as electrically isolated systems.

Interaction laws of the type of equations (1.2)–(1.3) have to contain at least two objects. Often one charge is of particular interest. Thus we may require the force on one charge due to many other charges. We know that there is a force of reaction on these other charges but we may not be particularly concerned with it. Suppose Q_2 is the charge of interest. We can write

$$\mathbf{F} = \mathbf{E}Q_2 \qquad (1.4)$$

where

$$\mathbf{E} = \frac{Q_1}{4\pi\varepsilon_0 r^2}\,\hat{\mathbf{r}}. \qquad (1.5)$$

This separates the interaction into the 'field' due to Q_1 which then acts on Q_2. More generally we can write

$$\mathbf{F} = \mathbf{E}Q \qquad (1.6)$$

where $\mathbf{E}$ is the field due to all the sources typified by Q_1 and Q is the charge of interest, which may be a test charge or probe with which we explore the field of the other charges. It is a small mental step to specify the problem entirely in terms of the electric field $\mathbf{E}$ because the probe is not an essential part of the system. Similarly we can write the energy as

$$U = VQ \qquad (1.7)$$

where V is the potential energy which would be associated with a charge of unit strength. This is defined as the potential at a point. The splitting of the interaction into two components has the enormous advantage mathematically of replacing quadratic by linear relationships. Thus both $\mathbf{E}$ and V are linearly dependent on the charges. Moreover, electrostatic systems generally employ conductors containing large amounts of free charge. This means that at equilibrium the electric field is zero inside the conductors which are therefore at constant potential. The problem is therefore generally specified in terms of potential rather than charge. Mathematically we then have a boundary value problem in which it is necessary to calculate the potential throughout a region subject to known values on the boundary.

A disadvantage inherent in the splitting of the interaction is that the potential is a point function and it requires much effort to find it at every point throughout a region. Happily such calculation schemes can be handled speedily by means of digital computers. However, there is also a conceptual disadvantage, and this may be more serious. The introduction of $\mathbf{E}$ and V has changed the point of view from that of interaction between material objects to that of action distributed in a field. We saw in §1.2 how these two views have been opposed over the centuries. We also saw that the two theories converged in the notion of a system combining matter and geometry. With regard to electrostatics we generally need the total energy of the system, whereas the potential gives the energy at a point. This implies that the region has to be explored with a minute probe, or if the probe is part of the system the charges must be subdivided into tiny droplets. By such means an energy and force distribution can be calculated with great accuracy. However, the question of whether the engineer needs to know this distribution is seldom asked. Generally he does not, but is content with total energy or with the force acting on a finite object like a conductor. If that is so, then the method of point potentials is uneconomical.

The interaction formulae of equations (1.1) and (1.2) are expressed entirely in terms of the sources without using a field description. Equations (1.6) and (1.7) are mixed and contain the field due to all the other sources acting on a particular point source. These mixed equations could be derived for any arbitrary law of interaction. However, the fact that this law varies inversely as the square of the distance enables us to develop expressions depending on field parameters only. Gauss's theorem gives rise to the concept of flux

$$\Psi = \sum Q \tag{1.8}$$

where

$$\Psi = \oint_s \mathbf{D} \cdot \mathbf{ds} \tag{1.9}$$

and the charges $\sum Q$ are in a region enclosed by the surface s. In free space the flux density $\mathbf{D} = \varepsilon_0 \mathbf{E}$, and in a polarizable material of relative permittivity ε_r, $\mathbf{D} = \varepsilon_0 \varepsilon_r \mathbf{E}$. This leads to an energy expression

$$U = \tfrac{1}{2} \int \mathbf{E} \cdot \mathbf{D} \, dv \tag{1.10}$$

and to forces which can be calculated from a stress system given by a normal tensile stress $\tfrac{1}{2} E_n D_n - \tfrac{1}{2} E_t D_t$ and a shear stress $E_t D_n$, where n and t denote the normal and tangential directions to the surface on which the stress acts. It should be noted that both the energy and the stress are quadratic in the field quantities, $\mathbf{E}$ is a measure of force per unit charge and $\mathbf{D}$ is a measure of quantity of charge derived through Gauss's theorem.

Electrostatic systems are by definition in a state of equilibrium. Any electrical force, as defined in equation (1.6), has therefore an equilibrating mechanical force opposing it. Such mechanical forces are ultimately also of electrical origin but they act at atomic and molecular distances. Electrostatics is concerned with distances that are large compared with atomic size; a reasonable minimum distance is of the order of 10^{-7} m, whereas atomic distances are of the order 10^{-10} m. Electric forces therefore describe a subsystem of the total system of forces. Inside a material both electrical and mechanical forces may be acting, and since only the total strain will be measurable the subdivision between electrical and mechanical action is somewhat arbitrary. However, whereas electrical forces and gravitational forces act at a distance, the close-range mechanical forces require material contact. This enables us to separate the types of force by summing the electrical force over a closed surface outside the body. The stress system

$$p = \tfrac{1}{2}E_n D_n - \tfrac{1}{2}E_t D_t$$
$$\tau = E_t D_n \qquad\qquad (1.11)$$

which has already been mentioned is therefore to be applied over a surface in free space. There are other stress systems which divide the electrical force between a surface stress and a body stress. As we have already noted this division is arbitrary and depends on the modelling of the structure of the material. For instance if we take electric dipoles as the smallest electric particles we shall obtain a different stress distribution from that which is based on the electric force on the ends of the dipoles. The reason is of course to be found in the existence of the additional internal mechanical force linking the ends of the dipoles. Only the sum of the surface and body forces has a unique value. It is therefore generally most convenient to use the free-space stress system of equation (1.11) which calculates the total force purely as a surface effect. Physically this means that the field is terminated at the surface and that all the sources have been transferred to that surface in such a manner as to leave the external field unchanged while reducing the internal field to zero. This is shown in Fig. 1.1. The normal field is terminated by a charge density $\sigma = -D_n$. In order to terminate the tangential field we must postulate a source which has the same effect on an electric field as the effect of an electric surface current on a magnetic field.

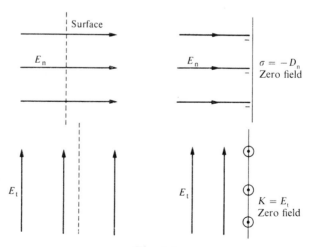

Fig. 1.1

We call this source a magnetic surface current. Equation (1.11) can be derived by noting that $p = \frac{1}{2}E_n\sigma - \frac{1}{2}D_tK$ and $\tau = \frac{1}{2}E_t\sigma + \frac{1}{2}D_nK$. It will be noted that the direction of K has been chosen as opposite to that of the electric current in the dual case of a magnetic field. This is done to preserve the difference of sign in the Maxwell equations for electromotive force and magnetomotive force.

These results in electrostatics are all derived from the inverse square law of force between electric charges. In the historical section we saw that this Newtonian law was applied by Michell to the interaction of magnetic poles. It was of course well known, and had been carefully explained by Gilbert, that isolated magnetic poles do not exist. The fundamental magnetic particles was therefore the dipole. However, the complicated law of force between dipoles could be greatly simplified by treating each pole separately. The existence of the theory of gravitation meant that results in magnetism, as in electricity, could be immediately derived from the Newtonian theory.

The dualism between electrostatics and magnetostatics is complete except for the one difference that the magnetic flux over a closed surface is zero. If pole strength is written as Q^* we have

$$\mathbf{F} = \frac{Q_1{}^* Q_2{}^*}{4\pi\mu_0 r^2}\,\hat{\mathbf{r}} \tag{1.12}$$

$$U = \frac{Q_1{}^* Q_2{}^*}{4\pi\mu_0 r} \tag{1.13}$$

$$\mathbf{F} = \mathbf{H}Q_2{}^* \tag{1.14}$$

where

$$\mathbf{H} = \frac{Q_1}{4\pi\mu_0 r^2}\,\hat{\mathbf{r}} \tag{1.15}$$

$$U = V^* Q^* \tag{1.16}$$

$$\Phi = \sum Q^* = 0 \tag{1.17}$$

where

$$\Phi = \oint \mathbf{B} \cdot d\mathbf{s} \tag{1.18}$$

$$U = \frac{1}{2} \int \mathbf{H} \cdot \mathbf{B} \, dv \tag{1.19}$$

$$\left.\begin{aligned} p &= \tfrac{1}{2}H_n B_n - \tfrac{1}{2}H_t B_t \\ \tau &= H_t B_n \end{aligned}\right\}. \tag{1.20}$$

So far the electric and magnetic systems are entirely separate, but we must now incorporate Oersted's and Ampère's experimental results which show that electric currents produce magnetic effects. Ampère's results can be summarized by the statement that the magnetic effect of a small current loop is identical to that of a small dipole. This equivalence can be extended to current loops of arbitrary size by Ampère's device of the magnetic shell. In this the actual loop of current is replaced by a surface distribution of small loops over an area bounded by the large loop. The currents in the small loops cancel everywhere except at the perimeter shown in Fig. 1.2. The equivalence is therefore complete except in the actual current-carrying loop itself. Outside a current-carrying region the sources of the magnetic field can be treated either as currents or as poles. If the magnetic interaction is due to currents we have a system of kinetic energy; if on the other hand the action is due to poles the energy is of the potential type. Since there is a distinction between these types of energy in mechanics, it is important to notice that this distinction arises from the specification of the sources and not from the field itself.

Since all magnetostatic effects can be handled by postulating current sources, whereas current-carrying regions must be excluded in considering the interaction of dipoles, it seems reasonable to treat currents as the more general type of source. Unfortunately this involves foregoing the simple structure of a mathematical formulation based on the inverse square law. Wherever possible we shall find it convenient to postulate magnetic poles as the sources of a system of potential energy.

The transference from potential to kinetic energy can be made as follows:

$$U = \tfrac{1}{2} \int_v \mathbf{H} \cdot \mathbf{B} \, dv. \tag{1.19}$$

Ampère's law can be written as

$$\mathbf{\nabla} \times \mathbf{H} = \mathbf{J} \tag{1.21}$$

Fig. 1.2

where $\mathbf{J}$ is the electric current density. Since $\mathbf{V} \cdot \mathbf{B} = 0$ we can introduce a new vector $\mathbf{A}$, where

$$\mathbf{V} \times \mathbf{A} = \mathbf{B} \tag{1.22}$$

and

$$\mathbf{V} \cdot \mathbf{A} = 0. \tag{1.23}$$

We have

$$U = \tfrac{1}{2} \int \mathbf{H} \cdot \mathbf{B} \, dv = \tfrac{1}{2} \int \mathbf{H} \cdot \mathbf{V} \times \mathbf{A} \, dv$$

$$= \tfrac{1}{2} \int \mathbf{A} \cdot \mathbf{V} \times \mathbf{H} \, dv + \tfrac{1}{2} \oint (\mathbf{A} \times \mathbf{H}) \cdot d\mathbf{s}. \tag{1.24}$$

For an isolated system the surface term can be incorporated in the volume by taking the surface just outside the system. We then have

$$U = \tfrac{1}{2} \int \mathbf{A} \cdot \mathbf{V} \times \mathbf{H} \, dv = \tfrac{1}{2} \int \mathbf{A} \cdot \mathbf{J} \, dv. \tag{1.25}$$

It only remains to find $\mathbf{A}$ in terms of the sources of the field. We have

$$\mathbf{V} \times \mathbf{A} = \mathbf{B} = \mu \mathbf{H}. \tag{1.26}$$

Hence

$$\mathbf{V} \times \mathbf{V} \times \mathbf{A} = \mathbf{V}(\mathbf{V} \cdot \mathbf{A}) - \nabla^2 \mathbf{A} = \mu \mathbf{J} \tag{1.27}$$

whence by the use of equation (1.23)

$$\nabla^2 \mathbf{A} = -\mu \mathbf{J} \tag{1.28}$$

and

$$\mathbf{A} = \int \frac{\mu \mathbf{J}}{4\pi r} \, dv. \tag{1.29}$$

The kinetic energy is now completely determined in terms of the current sources. It will be noted that the vector $\mathbf{A}$ gives an alternative description of the magnetic field. The energy density $\mathbf{A} \cdot \mathbf{J}$ has replaced $\mathbf{H} \cdot \mathbf{B}$. The kinetic energy density is therefore associated with the region occupied by the current sources, whereas the potential energy density $\mathbf{H} \cdot \mathbf{B}$ is associated with all the space occupied by a magnetic system. Physically only the total energy has any meaning. In both cases it is the mutual energy of the sources which is under consideration. The distribution of this energy depends on the models chosen for the sources. The physical meaning of the vector $\mathbf{A}$ can be clarified by examining the energy density $\mathbf{A} \cdot \mathbf{J}$. The current density $\mathbf{J}$ is due to the motion of a charge density ρ with velocity $\mathbf{v}$. We have $\mathbf{J} = \rho \mathbf{v}$. The kinetic energy density is therefore $\rho \mathbf{v} \cdot \mathbf{A}$. Thus $\mathbf{A}$ is a momentum per unit charge. It is important to notice that this is not the momentum of the local charge density ρ but the momentum of the entire current system described by equation (1.29). In non-relativistic contact mechanics kinetic energy is entirely a local phenomenon. In electromagnetism there is action at a distance, and kinetic energy as well as potential energy are mutual effects belonging to the system as a whole. This explains why the energy can be calculated equally well by equation (1.19) and equation (1.25). The vector $\mathbf{A}$ is generally called the vector potential, a somewhat unfortunate nomenclature since potential energy is necessarily a scalar. It might have been better to adopt Maxwell's term 'electrokinetic momentum'.

We now turn to Faraday's law which complements Ampere's law, given in equation (1.21), by describing the electric effect of a changing magnetic field. We have

$$\mathbf{V} \times \mathbf{E} = -\frac{\partial \mathbf{B}}{\partial t}. \tag{1.30}$$

On substituting the vector $\mathbf{A}$ for $\mathbf{B}$ by means of equation (1.22) and then integrating we obtain

$$\mathbf{E} = -\frac{\partial \mathbf{A}}{\partial t} - \mathbf{V}V. \tag{1.31}$$

Thus the force per unit charge $\mathbf{E}$ equals the negative rate of change of momentum per unit charge plus the negative gradient of potential energy per unit charge. This equation is strictly analogous to the mechanical force equation, the only change being that charge has been substituted for mass.

To complete this brief review it is necessary to mention Maxwell's criticism of Ampere's law applied to time-varying effects. The equation of continuity of charge can be written

$$\mathbf{V} \cdot \mathbf{J} + \frac{\partial \rho}{\partial t} = 0. \tag{1.32}$$

This contradicts equation (1.21) which leads to

$$\mathbf{V} \cdot \mathbf{J} = 0, \tag{1.33}$$

a statement which is true for steady currents only. Maxwell modified Ampère's law to the more general form

$$\mathbf{V} \times \mathbf{H} = \mathbf{J} + \frac{\partial \mathbf{D}}{\partial t}. \tag{1.34}$$

By the use of equation (1.22)

$$\mathbf{V} \times \mathbf{B} = \mathbf{V} \times \mathbf{V} \times \mathbf{A} = \mu \mathbf{J} + \mu \varepsilon \frac{\partial \mathbf{E}}{\partial t} \tag{1.35}$$

and by the use of equation (1.31)

$$\mathbf{V} \times \mathbf{V} \times \mathbf{A} = \mathbf{V}(\mathbf{V} \cdot \mathbf{A}) - \mathbf{V}^2\mathbf{A}$$

$$= \mu\mathbf{J} - \mu\varepsilon\frac{\partial^2\mathbf{A}}{\partial t^2} - \mu\varepsilon\mathbf{V}\frac{\partial V}{\partial t}. \tag{1.36}$$

If the static definition of the divergence of **A** in equation (1.23) is replaced by the more general statement

$$\mathbf{V} \cdot \mathbf{A} + \mu\varepsilon\frac{\partial V}{\partial t} = 0 \tag{1.37}$$

and if $\mu\varepsilon$ is replaced by $1/c^2$, where c is the velocity of light, we have

$$\mathbf{V}^2\mathbf{A} - \frac{1}{c^2}\frac{\partial^2\mathbf{A}}{\partial t^2} = -\mu\mathbf{J} \tag{1.38}$$

which should be compared with the magnetostatic equation (1.28). It leads to the integral expression

$$\mathbf{A} = \int\frac{\mu[\mathbf{J}]}{4\pi r}\,dv \tag{1.39}$$

The expression [**J**] implies that the value of the current density **J** is to be taken at the earlier time $t - r/c$. **A** is known as the delayed vector potential. By the use of

$$\mathbf{V} \cdot \mathbf{D} = \rho \tag{1.40}$$

a parallel expression can be derived for the scalar potential V:

$$\mathbf{V}^2 V - \frac{1}{c^2}\frac{\partial^2 V}{\partial t^2} = -\frac{\rho}{\varepsilon} \tag{1.41}$$

and

$$V = \int\frac{[\rho]}{4\pi\varepsilon r}\,dv. \tag{1.42}$$

It will be noted that the relationship between **A** and V given in equation (1.37) is based on the equation of continuity (1.32) and has the desirable effect that it makes **A** depend only on **J** and V only on ρ. The distinction between kinetic and potential energy is therefore preserved. It will also be noted that Maxwell's hypothesis of equation (1.34) converts the instantaneous interactions into effects which are propagated with the finite velocity c. This forms the basis of relativistic mechanics.

The relative importance of the time delay r/c depends both on the time scale, and therefore frequency, of the phenomenon and on the distances. A rough measure is the ratio of distance to electromagnetic wavelength. If this ratio is greater than unity the time delay is crucial, but if it is smaller the delayed potentials have values which are near those for static fields. If it is remembered that the wavelength at 50 Hz is 6×10^6 m, it follows that there is a large range of devices for which the time delay is unimportant.

Further reading

A superb account of the history of the subject is given in *A history of the theories of aether and electricity, Vol. 1, The classical theories,* by Sir Edmund Whittaker, Nelson, London (1951).

Greek scientific thought is presented lucidly and concisely in *The physical world of the Greeks,* by S. Sambursky, Routledge, London (1959).

Interesting historical details are given in *Biographical history of electricity and magnetism,* by P. F. Mottelay, Griffin (1922).

A more leisurely account of the content of §1.3 is given in many books on electromagnetic theory including *Applied electromagnetism,* by P. Hammond, Pergamon Press, Oxford (1971).

2. Variational Principles in Mechanics

2.1. The principle of virtual work

Our historical introduction has highlighted two different kinds of physical theory, so it is not surprising that there are also two types of mechanics. The more familiar of the two starts with Newton's laws of motion which are based on the idea of mass points or particles. If the force on a particle is prescribed, its motion can be obtained by integration of the differential equation known as Newton's second law. The development of this type of mechanics arose in the context of the motion of the planets under the action of gravity. In these problems the planets could be idealized as particles because of the large distances between them. The scheme of calculation involves finding the vector forces acting on the particles. The study of electric and magnetic interactions based on the inverse square law followed the same Newtonian scheme, and the determination of the electric and magnetic field involves finding the force at every point in a region of space. Clearly this method is ideal for systems of a few particles but it is cumbersome for large assemblages. Electromagnetic systems generally consist of distributions of charge, current, and pole strength, the actual location of these sources being unspecified. We may for example know the total charge on a conductor but are likely to be ignorant of its distribution. All that we can say about it is that the particles will so dispose themselves as to be in equilibrium. Similar difficulties were first encountered in considering the mechanics of systems, and they were surmounted by a method known as analytical mechanics which is associated with the name of the great French mathematician Lagrange (1736–1813). Analytical mechanics lies in the historical succession of 'plenum' or 'continuum' theories as against 'action at a distance' or 'vacuum' theories. Of course these distinctions should not be regarded in terms of conflict but rather in terms of a different angle or viewpoint. Our brief treatment is based on the brilliant book by C. Lanczos† to which the reader should refer for a full account. (Readers who are familiar with variational methods in mechanics can omit this chapter

† C. Lanczos, *The variational principles of mechanics*, University of Toronto Press, Toronto (1974).

and proceed at once to the application of variational methods to electromagnetism discussed in Chapter 3.) Analytical mechanics has as its motive the description of complete systems in terms of invariant parameters, foremost amongst which is the energy of the system. This not only reduces the number of entities to be considered but also disentangles the physical phenomena from the structure of arbitrary systems of co-ordinates because the parameters are invariant under suitable transformations of co-ordinates. The study of such transformations, which is an important aspect of analytical mechanics, has contributed greatly to the understanding of the physical processes themselves and is the basis of relativity theory.

Whereas the starting point of particle mechanics is force, analytical mechanics starts with work. We consider the infinitesimal work done by the impressed forces as

$$\delta w = F_1 \, \delta q_1 + F_2 \, \delta q_2 + \ldots + F_n \, \delta q_n \tag{2.1}$$

where $F_1, F_2, \ldots, F_n$ are the components of force associated with the displacement of the co-ordinates $q_1, q_2, \ldots, q_n$. These co-ordinates are known as the generalized co-ordinates and they can be expressed in terms of any arbitrary set of co-ordinates, rectangular or curvilinear. The transformation of co-ordinates is subject to the invariance of the magnitude of the line element δs which determines distances in the configuration space. The number of generalized co-ordinates is the number of degrees of freedom of the system. This number is of course likely to be very much less than the number of particles of a system because the particles will in general have constraints on their motion imposed by connections of various kinds. The inclusion of these connections in the co-ordinate system is one of the great advantages of this method of mechanics. The force associated with the generalized co-ordinates is known as the generalized force and also has as many components as the number of degrees of freedom of the system.

If the expression of equation (2.1) is integrable we can connect it with a potential energy $U(q_1, q_2, \ldots, q_n)$ such that

$$\sum F_i \, dq_i = -\sum \frac{\partial U}{\partial q_i} \, dq_i \tag{2.2}$$

and the components of the generalized force can then be derived from the single function U by differentiation:

$$F_i = -\frac{\partial U}{\partial q_i} \tag{2.3}$$

This is possible in static systems which satisfy the law of conservation of energy. The method can, however, be extended to dynamic systems and it is one of the chief objects of analytical mechanics to develop this possibility using as the guiding principle the notion of equilibrium.

In Newtonian mechanics equilibrium requires that the resultant forces and torques on a body shall be zero. In Lagrangian mechanics these conditions are incorporated in the principle of virtual work which states that for equilibrium

$$\delta w \leqslant 0. \tag{2.4}$$

In the special case described by equation (2.3) we have

$$\delta U \geqslant 0, \tag{2.5}$$

but the principle of virtual work is not confined to static conservative systems. If the displacements are reversible the relationships (2.4) and (2.5) lose the inequality signs and the virtual work is zero at equilibrium. The inequality specifies the condition that displacement is impossible in some direction if the system is on the boundary of its configuration space. A typical example is a weight suspended from a string which cannot be displaced vertically downwards.

If the system is characterized by a potential energy U the principle of virtual work states that at equilibrium a possible variation of this energy is greater than or equal to zero. The energy is therefore a minimum. Instead of considering the equilibrium of forces we are led to consider the variation of a single energy parameter. Attention is transferred from the parts of a system to the system as a whole.

So far we have considered static systems and we now seek to include dynamics in the same procedure. The key to this development is provided by d'Alembert's principle of reversed effective forces. This replaces Newton's law

$$\mathbf{F} = m\mathbf{a} \tag{2.6}$$

by

$$\mathbf{F} - m\mathbf{a} = 0 \tag{2.7}$$

where $-m\mathbf{a}$ is the force of inertia. This apparently trivial device produces a profound change of viewpoint. Dynamics is incorporated in statics and accelerated motion becomes a system concept. We now enlarge the statement of equation (2.1) by writing for a system of particles

$$\delta w = \sum \mathbf{F}_i \cdot \delta \mathbf{R}_i - \sum \frac{\mathrm{d}}{\mathrm{d}t}(m_i \mathbf{v}_i) \cdot \delta \mathbf{R}_i. \tag{2.8}$$

where $\mathbf{F}_i$ represents the given external forces and $\delta \mathbf{R}_i$ the corresponding virtual displacements. Equation (2.8) is a special case of the general statement of equation (2.1). If the forces $\mathbf{F}_i$ can be derived from a potential energy U we can write

$$\delta w = -\delta U - \sum \frac{\mathrm{d}}{\mathrm{d}t}(m_i \mathbf{v}_i) \cdot \delta \mathbf{R}_i. \tag{2.9}$$

This statement can be simplified if we integrate with respect to time between the limits t_1 and t_2:

$$\int_{t_1}^{t_2} \delta w \, dt = -\int_{t_1}^{t_2} \delta U \, dt - \sum \int_{t_1}^{t_2} \frac{d}{dt}(m_i \mathbf{v}_i \cdot \delta \mathbf{R}_i) \, dt$$

$$+ \sum \int_{t_1}^{t_2} m_i \mathbf{v}_i \cdot \frac{d}{dt}(\delta \mathbf{R}_i) \, dt. \tag{2.10}$$

Since $(d/dt)(\delta \mathbf{R}_i) = \delta \mathbf{v}_i$ we obtain

$$\int_{t_1}^{t_2} \delta w \, dt = \int_{t_1}^{t_2} \delta \tfrac{1}{2} \sum m_i \mathbf{v}_i^2 \, dt - \int_{t_1}^{t_2} \delta U \, dt$$

$$- \left[\sum m_i \mathbf{v}_i \cdot \delta \mathbf{R}_i \right]_{t_1}^{t_2}. \tag{2.11}$$

The variational δ can be interchanged with the integration because the processes are entirely independent. If we define the kinetic energy by

$$T = \tfrac{1}{2} \sum m_i v_i^2 \tag{2.12}$$

we obtain

$$\int_{t_1}^{t_2} \delta w \, dt = \delta \int_{t_1}^{t_2} T \, dt - \delta \int_{t_1}^{t_2} U \, dt$$

$$- \left[\sum m_i \mathbf{v}_i \cdot \delta \mathbf{R}_i \right]_{t_1}^{t_2}. \tag{2.13}$$

Finally we prescribe that there shall be no variations of the system configuration at the times $t = t_1$ and $t = t_2$ which define the beginning and end of the motion, so that the variation is carried out between prescribed limits. This removes the boundary term and we have

$$\int_{t_1}^{t_2} \delta w \, dt = \delta \int_{t_1}^{t_2} (T - U) \, dt = \delta \int_{t_1}^{t_2} L \, dt \tag{2.14}$$

where $L = T - U$ is the Lagrangian energy. The definite integral of the virtual work with respect to time becomes the variation of the definite integral

$$A = \int_{t_1}^{t_2} L \, dt \tag{2.15}$$

which is called the action.

Now d'Alembert's principle requires δw to varnish at any time. Hence it can be written as

$$\delta A = 0. \tag{2.16}$$

This is known as Hamilton's principle and focuses attention on the system by means of the single action integral. The Lagrangian L and the action A are invariant under transformations of the general co-ordinates q_i.

The necessary and sufficient conditions for the stationary value of A given in equation (2.16) are given by the Euler–Lagrange differential equations (see the Appendix to this chapter):

$$\frac{d}{dt} \frac{dL}{\partial \dot{q}_i} - \frac{\partial L}{\partial q_i} = 0, \qquad (i = 1, 2, \ldots, n) \tag{2.17}$$

where the $\dot{q}_i$ are the velocities associated with the co-ordinates q_i. Under a transformation of the co-ordinates the individual equations for the co-ordinates will be changed but the complete set of equations is invariant and can be used to describe the equations of motion in any co-ordinate system.

The inequality in equation (2.4) which describes the equilibrium of a system on the boundary of its configuration space implies that

$$\delta(T - U) \leqslant 0. \tag{2.18}$$

Hence for a purely kinetic system T has a maximum stationary value which should be compared with the minimum stationary potential energy U in a static system. This difference in the behaviour of T and U provides a useful double bound in numerical calculations and will be discussed later in this chapter and in subsequent chapters. It arises from d'Alembert's principle and the subsequent integration of the work of the reversed effective forces.

2.2. Lagrange's multipliers

The Euler–Langrange equations (2.17) are a group of n equations for a system of n degrees of freedom. The connections between the parts of a system reduce the number of degrees of freedom which the system would have without those connections. For example a bead might be constrained to move along a wire. Such constraints could be embodied in the equations by eliminating some of the variables. For instance a system with m constraints would have $n - m$ degrees of freedom and $n - m$ independent variables. Elimination may be tedious and in any case the distinction between the independent and the dependent variables may be artificial. For these reasons it is useful to adopt instead the elegant method devised by Lagrange and known as the method of undetermined multipliers.

Suppose we have a function

$$F = F(q_1, q_2, \ldots, q_n) \tag{2.19}$$

subject to a constraint

$$f = f(q_1, q_2, \ldots, q_n) = 0. \tag{2.20}$$

We are interested in the variation of F subject to the condition $f = 0$. For a stationary value of F we require

$$\delta F = \frac{\partial F}{\partial q_i} \delta q_1 + \ldots + \frac{\partial F}{\partial q_n} \delta q_n = 0. \tag{2.21}$$

From (2.20) we have

$$\delta f = \frac{\partial f}{\partial q_i} \delta q_i + \ldots + \frac{\partial f}{\partial q_n} \delta q_n = 0. \tag{2.22}$$

Let

$$\lambda = \lambda(q_1, q_2, \ldots, q_n) \tag{2.23}$$

be an undetermined function of the qs. We can write

$$\delta F + \lambda \, \delta f = 0. \tag{2.24}$$

Suppose that we use equation (2.20) to eliminate one of the variables q. It could be any of the n variables, but we shall choose the last one, q_n. This can be done by rewriting equation (2.24) as

$$\sum_{i=1}^{n} \left(\frac{\partial F}{\partial q_i} + \lambda \frac{\partial f}{\partial q_i} \right) \delta q_i = 0 \tag{2.25}$$

and choosing λ such that the term in δq_n is zero, i.e.

$$\frac{\partial F}{\partial q_n} + \lambda \frac{\partial f}{\partial q_n} = 0. \tag{2.26}$$

We now have a problem in $n - 1$ independent variables for which

$$\sum_{i=1}^{n-1} \left(\frac{\partial F}{\partial q_i} + \lambda \frac{\partial f}{\partial q_i} \right) \delta q_i = 0. \tag{2.27}$$

This means that each coefficient

$$\frac{\partial F}{\partial q_i} + \lambda \frac{\partial f}{\partial q_i} = 0 \qquad (i = 1, 2, \ldots, n-1). \tag{2.28}$$

Hence combining (2.28) with (2.26) we have a system of n relationships. The introduction of λ has enabled us to treat all the q_i equally, but instead of considering $\delta F = 0$ we consider $\delta F + \lambda \, \delta f = 0$. Alternatively we can consider

$$\delta(F + \lambda f) = \delta F + \delta \lambda f + \lambda \, \delta f = 0 \tag{2.29}$$

since the constraint is $f = 0$ by equation (2.20). Thus we replace the original function F by the function $F + \lambda f$. Clearly this can be generalized for a system having m constraints ($m < n$). We then examine the variation

$$\delta(F + \lambda_1 f_1 + \ldots + \lambda_m f_m) = 0. \tag{2.30}$$

By this means the constraints, or auxiliary conditions, can be inserted into the function to be varied without having recourse to elimination. All the initial co-ordinates q are treated equally.

This remarkable method can be applied to the variation of integrals such as the action integral of equation (2.15). Instead of the set of equations (2.17) we have (see the Appendix to this chapter),

$$\frac{d}{dt}\frac{\partial L}{\partial q_i} - \frac{\partial L}{\partial q_i} - \lambda_1 \frac{\partial f_1}{\partial q_i} - \lambda_2 \frac{\partial f_2}{\partial q_i} - \ldots - \lambda_m \frac{\partial f_m}{\partial q_i} = 0. \tag{2.31}$$

The Lagrangian function L is replaced by the function L' where

$$L' = L - \lambda_1 f_1 - \ldots - \lambda_m f_m. \tag{2.32}$$

The sign of the λs is arbitrary, but since $L = T - U$ we are associating the extra terms with U rather than T. With this substitution the system can again be treated as having n degrees of freedom. Instead of eliminating m variables and solving $n - m$ equations we introduce m undetermined multipliers and increase the number of variables to $n + m$. However, the m constraints are given, so that the augmented system has once again n variables.

The reader will have guessed that the Lagrangian multipliers are 'undetermined' only in a formal sense, but that they have a very clear physical interpretation. We are discussing the motion of a mechanical system. Into such a system having n degrees of freedom there are introduced certain kinematical constraints in terms of relationships between the co-ordinates. Mechanically these constraints are associated with the forces which maintain them and these forces are associated with the Lagrangian multipliers. Instead of introducing the constraints by elimination we allow a free variation of all the co-ordinates. This means that in the principle of virtual work we must include the virtual work terms $\Sigma \lambda \, \delta f$ of the forces maintaining the constraints. The term $\Sigma \lambda f$ represents an additional potential energy and the forces can be obtained by differentiation. Thus the force associated with λ and the co-ordinate q_i is given by

$$F_{1_i} = \frac{\partial(\lambda_1 f)}{\partial q_i} = -\lambda_1 \frac{\partial f}{\partial q_i} - \frac{\partial \lambda_1}{\partial q_i} f \tag{2.33}$$

However, $f = 0$ by equation (2.20) so that

$$F_{1_i} = -\lambda_1 \frac{\partial f}{\partial q_i} \tag{2.34}$$

and the Lagrangian multiplier provides the force of reaction of the constraint. Physically this means that the constraint has to alter microscopically to provide the reaction force. The smaller the variation, the greater has to be the force. It will also be noted that action and reaction cancel out, as required by Newton's third law or the principle of virtual work, so that we do not need to introduce the reaction forces nor the Lagrangian multipliers. This can be done by using the constraints as a means of reducing the independent variables by elimination.

2.3. Conservation of energy

In particle mechanics and in the mechanics of rigid bodies the conditions of equalibrium are obtained by considering the forces and torques. Each part of a system then has its own equilibrium equations and the systems as a whole can be treated by eliminating the reactions between the different members. Displacements are treated separately by the equations of compatibility.

   In system mechanics, on the other hand, equilibrium conditions are derived from the principle of virtual work which combines force and displacement and concentrates on energy. Hamilton's principle $\delta A = 0$ (2.16) shows how the equilibrium conditions can be obtained by considering the variation of a scalar quantity, the action, which is the definite integral of the Lagrangian function $L = T - U$ where T and U are the kinetic and potential energies respectively. In this treatment the idea of equilibrium has been extended to include dynamic as well as static systems by means of d'Alembert's principle of reversed effective forces. Thus the programme of system mechanics is to find the stationary value of the action and therefore of the Lagrangian energy difference. Auxiliary conditions, such as kinematic constraints, are built into the Lagrangian at the outset by means of the method of Lagrangian multipliers. The force equations can be derived from the variational principle, but it is a profound mistake to treat the method as a means of finding these equations which can often be derived more easily by inspection. The method seeks the system parameters at equilibrium rather than the equations of motion, and in this lies its great advantage. Once these parameters have been found the motion of the whole system is determined.

Here, however, we have to face a difficulty. It will have been noted that Hamilton's principle was derived by integrating the virtual work. This involved the assumption that the applied forces could be derived from an energy function and hence from a potential energy. The reversed effective forces were then derived from a kinetic energy. It is clear that not all applied forces can be derived from a potential energy. There may be other forms of energy or energy may be interchanged with the environment. For example there may be frictional forces present. If this is so we cannot in general integrate the virtual work.

We note from the form of the Euler–Lagrange equation (2.17) that the most general form of a force derivable from a work function is

$$F_i = -\frac{\partial U}{\partial q_i} + \frac{\mathrm{d}}{\mathrm{d}t}\frac{\partial U}{\partial \dot{q}_i}. \qquad (2.35)$$

If U is independent of the velocities $\dot{q}_i$, this equation reduces to the familiar form

$$F_i = -\frac{\partial U}{\partial q_i} \qquad (2.36)$$

where U can now be identified with the potential energy.

If the applied forces are derivable from the potential energy as shown in equation (2.36) and if the masses of the particles are assumed constant, d'Alembert's form of the principle of virtual work can be written

$$\delta U + \sum m_i \frac{\mathrm{d}}{\mathrm{d}t}(\mathbf{v}_i) \cdot \delta \mathbf{R}_i = 0. \qquad (2.37)$$

Let us now restrict the arbitrary variations $\delta \mathbf{R}_i$ to being the *actual* displacements during a time $\mathrm{d}t$. For this special variation δU becomes the actual change of potential energy during the time interval and we shall replace the δs by ds and write

$$\mathrm{d}U + \sum m_i \ddot{\mathbf{R}}_i \cdot \mathrm{d}\mathbf{R}_i = \mathrm{d}U + \sum m_i \ddot{\mathbf{R}}_i \cdot \mathrm{d}\dot{\mathbf{R}}_i \, \mathrm{d}t$$

$$= \mathrm{d}U + \frac{\mathrm{d}}{\mathrm{d}t}(\tfrac{1}{2}\sum m_i \dot{R}_i{}^2)\,\mathrm{d}t$$

$$= \mathrm{d}U + \mathrm{d}T = 0 \qquad (2.38)$$

where $T = \frac{1}{2} \Sigma\, m_i \dot{R}_i{}^2$ is defined as the kinetic energy. Equation (2.38) can be integrated with respect to time and we obtain for the motion

$$U + T = E \qquad (2.39)$$

where E is the total energy of the system expressed as the sum of the potential and the kinetic energy. The principle of the conservation of energy therefore applies to this system.

Let us now consider this conservation law in the more general context of Hamilton's principle. In order to do this we must relax the condition that the co-ordinates are not varied at the end-points in time. This must be done so that we can consider the actual displacements dq_i, which are dependent on the velocities, whereas the arbitrary δq_i need not be so.

From the calculus of variations we have (see equation (2.101) in the Appendix to this chapter)

$$\delta \int_{t_1}^{t_2} L \, dt = \left[\sum \frac{\partial L}{\partial \dot{q}_i} \delta q_i \right]_{t_1}^{t_2}. \qquad (2.40)$$

The boundary term on the right-hand side has previously been deleted by prescribing $\delta q_i = 0$ at $t = t_1$ and $t = t_2$. We shall put

$$p_i = \frac{\partial L}{\partial \dot{q}_i} \qquad (2.41)$$

both in order to simplify the notation and also because we shall discuss the nature of this function in the next section when we deal with the generalized momenta as well as the co-ordinates and velocities. We now require that L shall not contain the time t explicitly, so that

$$L = L(q_1, \ldots, q_n; \dot{q}_1, \ldots, \dot{q}_n). \qquad (2.42)$$

Now since we are dealing with the actual variations during the short interval $dt = \varepsilon$, we can write

$$\delta L = dL = \varepsilon \dot{L} \qquad (2.43)$$

and

$$\delta q_i = dq_i = \varepsilon \dot{q}_i. \qquad (2.44)$$

Hence equation (2.40) becomes

$$\delta \int_{t_1}^{t_2} L \, dt = \int_{t_1}^{t_2} \delta L \, dt$$

$$= \varepsilon [L]_{t_1}^{t_2} = \varepsilon \left[\sum p_i \dot{q}_i \right]_{t_1}^{t_2}. \qquad (2.45)$$

Now t_1 and t_2 do not enter into the equation of motion and are arbitrary. Therefore, in order to satisfy equation (2.45) we have

$$\sum p_i \dot{q}_i - L = \text{constant}. \tag{2.46}$$

If the kinetic energy T is a quadratic form of the velocities

$$T = \tfrac{1}{2} \sum_{i,k=1}^{n} a_{ik} \dot{q}_i \dot{q}_k \tag{2.47}$$

and if U is independent of the velocities

$$p_i = \frac{\partial L}{\partial \dot{q}_i} = \frac{\partial T}{\partial \dot{q}_i} = \sum_{R=1}^{n} a_{ik} \dot{q}_k. \tag{2.48}$$

Then

$$\sum_{i=1}^{n} p_i \dot{q}_i = 2T \tag{2.49}$$

and equation (2.45) becomes

$$T + U = \text{constant} = E. \tag{2.50}$$

Thus the law of conservation of energy in the form of the constancy of kinetic plus potential energy is true under certain conditions. These are that L does not depend on the time explicitly, that U is independent of the velocities, and that T is a quadratic form of the velocities.

The last requirement is not met if there are gyroscopic terms in the expression for the energy. Such terms are linear in the velocities. The magnetic force of a current on a moving charge is of this kind. The condition that U is independent of the velocities is normally true, but does not apply to relativistic mechanics. Without these two restrictions equation (2.50) does not hold, but equation (2.46) can then be used as a more general conservation law where the constant is defined as the total energy of the system. This law depends on the condition that L has the form of equation (2.42).

We conclude that Hamilton's principle is more general than the law of the conservation of energy in its usual form but it requires that the applied forces can be derived from a potential function U. The integration of the virtual work of the reversed effective forces with respect to time results in the definition of a further term, the kinetic energy, and so completes the integration and establishes Hamilton's principle.

The distinction between systems in which L is explicitly dependent or independent of time determines whether the arbitrary variations δq_i can be identified with the real variations dq_i in time dt. An explicit time dependence connects the displacements with the velocities and restricts the variations which are possible. The same difficulty arises when the constraints or auxiliary conditions are explicitly dependent on time. A similar restriction holds if the kinematic constraints are given in 'non-holonomic' form. A holonomic constraint is defined as a relationship between the co-ordinates q_i. From this relationship variations δq_i can be obtained by differentiation but the reverse process may not be possible. It may be that the constraint is given in terms of velocities or infinitesimal displacements; then it may not be possible to integrate the relationship to obtain an equation in terms of the co-ordinates. The example which is generally quoted is the rolling of a sphere on a plane without slipping. In this motion there are two degrees of freedom. If the point of contact has co-ordinates x and y and if these are given in terms of the time t, the motion is determined. However, the general turning motion of a sphere needs three angular co-ordinates, say α, β, and γ. The constraint of rolling on a plane can be expressed in terms of relationships between the differentials of α, β, and γ and of x and y. However, these relationships cannot be integrated. Instead of having the condition

$$f(q_1, q_2, \ldots, q_n) = 0 \qquad (2.20)$$

we have

$$A_1 \, dq_1 + A_2 \, dq_2 + \ldots + A_n \, dq_n = 0 \qquad (2.51)$$

where the coefficients A_i are functions of the q_i.

It might seem that the difficulties caused by non-conservative systems and non-holonomic constraints severely restrict the usefulness of Lagrangian mechanics. This, however, is not so. First many systems do obey the conservation of energy and secondly non-holonomic constraints and auxiliary conditions depending on time explicitly are rare. Moreover, even these difficulties can be overcome by various stratagems. For a full account of these the interested reader should refer to a book like that by Lanczos which has already been mentioned. The most obvious method is to retain the principle of virtual work in its differential form. When this is done the non-conservative forces can be segregated and so can the non-holomic constraints. The Lagrangian energy retains its importance as a system parameter although there may be other terms contributing to the virtual work equation. The method of Lagrange's multipliers still retains its value in dealing with constraints and auxiliary conditions. All that has to be done is to use it in the form $\delta F + \lambda \, \delta f$ as in equation (2.24) instead of $F + \lambda f$ as in equation (2.29).

2.4. Co-ordinates and momenta

Hamilton's principle considers the variation of the Lagrangian energy L which is a function of the generalized co-ordinates q_i and velocities $\dot{q}_i$. The equations of motion, or of equilibrium if we think in terms of d'Alembert's reversed effective forces, are a set of n differential equations of the second order. Hamilton discovered a transformation which converts the n second-order equations of motion to $2n$ first-order differential equations and at the same time causes the energy to be linear in the velocities instead of being quadratic. The basis of this transformation is the idea of momentum. We know that in Newtonian mechanics we can replace the product of mass and velocity by momentum. This is particularly useful in systems of varying mass and we shall find it equally important in systems of varying charge, but these are incidental benefits. The overriding importance of the transformation from velocity to momentum lies in the change of viewpoint. We have already discussed how the apparently trivial transfer of the effective acceleration forces in d'Alembert's principle from the right-hand side of the force equation to the left-hand side at one stroke converts dynamics into a branch of statics. Hamilton's transformation is as revolutionary. It shows that the momenta can be treated as another set of co-ordinates, and thus introduces a beautiful duality and symmetry which not only simplifies the theory but also forms the basis of the numerical methods with which we shall deal in later chapters.

The mathematical basis of the method can best be introduced by a discussion of the Legendre transformation of co-ordinates. Suppose we have a function of two variables $f(x, y)$ and that therefore for an arbitrary variation

$$\delta f = u\,\delta x + v\,\delta y \tag{2.52}$$

where

$$u = \frac{\partial f}{\partial x} \quad \text{and} \quad v = \frac{\partial f}{\partial y}.$$

It is desired to change the description from the variables x, y to new independent variables u, y. Let us introduce a new function $g(u, y)$ where

$$\delta g = \frac{\partial g}{\partial u}\,\delta u + \frac{\partial g}{\partial y}\,\delta y \tag{2.53}$$

and let the relationship between g and f be given by the Legendre transformation

$$g = ux - f \tag{2.54}$$

so that

$$\delta g = u\,\delta x + x\,\delta u - \delta f \tag{2.55}$$

whence by equation (2.52)

$$\delta g = x\,\delta u - u\,\delta y \tag{2.56}$$

and by comparison with equation (2.53)

$$x = \frac{\partial g}{\partial u}, \qquad u = -\frac{\partial g}{\partial y}. \qquad (2.57)$$

In this transformation from f to g we have transformed the 'active' variable x to the new variable u while the 'passive' variable y has been retained. We notice that the Legendre transformation is symmetrical, so that one could write

$$f = xu - g \qquad (2.58)$$

and consider u as the active variable to be eliminated in favour of x. Thus we have the pair of relationships

$$x = \frac{\partial g}{\partial u}, \qquad u = \frac{\partial f}{\partial x}. \qquad (2.59)$$

We also note that for the passive variables

$$\frac{\partial f}{\partial y} = -\frac{\partial g}{\partial y}. \qquad (2.60)$$

This could of course be written

$$\frac{\partial g}{\partial y} = -\frac{\partial f}{\partial y} \qquad (2.61)$$

so that the symmetry of the transformation is complete.

We now apply these ideas to the Lagrangian function $L\ (q_1, \ldots, q_n; \dot{q}_1, \ldots, \dot{q}_n; t)$. We seek to eliminate the velocities $\dot{q}_i$ by treating them as the active variables in a Legendre transformation. These velocities are to be replaced by the generalized momenta already defined (2.41) as

$$p_i = \frac{\partial L}{\partial \dot{q}_i} \qquad (2.41)$$

and it will be noted that this is in accordance with equation (2.59) where the function f has been replaced by L. The new function corresponding to g is written $H\ (q_1, \ldots, q_n; p_1, \ldots, p_n; t)$ and therefore from equation (2.54)

$$H = \sum_{i=1}^{n} p_i \dot{q}_i - L. \qquad (2.62)$$

From equation (2.59) we see that we can write

$$\dot{q}_i = \frac{\partial H}{\partial p_i}. \qquad (2.63)$$

The Lagrangian equations of motion (2.17) can be written by the use of the definition of equation (2.41) as

$$\dot{p}_i = \frac{\partial L}{\partial q_i}.$$ (2.64)

By the use of the relationship (2.60) for the passive variables we can write

$$\dot{p}_i = -\frac{\partial H}{\partial q_i}.$$ (2.65)

The equation of motion in terms of the new Hamiltonian function H are therefore given in equations (2.63) and (2.65). We have therefore replaced the n second-order equations (2.17) by $2n$ first-order equations. Mathematically the q_i and p_i have the same status and can be interchanged if desired.

The action integral can now be written from equations (2.62) and (2.15) as

$$A = \int_{t_1}^{t_2} \left\{ \sum p_i \dot{q}_i - H(q_1, \ldots, q_n; p_1, \ldots, p_n; t) \right\} dt.$$ (2.66)

Previously the variation of this integral and the variation of the integrand L were considered in terms of variations of the q_i and $\dot{q}_i$. The new integrand has in addition the momenta p_i. These are not independent of the q_i and $\dot{q}_i$ and it might seem that before performing the variation they need to be eliminated. However, this is unnecessary for the following reason.

Consider an arbitrary variation of the p_i:

$$\delta L = \sum_{i=1}^{n} \left[\dot{q}_i - \frac{\partial H}{\partial p_i} \right] \delta p_i.$$ (2.67)

The term in parentheses is zero by equation (2.63) and so the variation of the p_i does not affect the stationary value of the action integral. This is a very interesting discovery which links the Legendre transformation with the method of the Lagrangian multipliers. In considering that method we noticed that it was possible to replace the statement $\delta F + \lambda \, \delta f = 0$ of equation (2.27) by the wider statement $\delta(F + \lambda f) = 0$ of equation (2.29). Reference to those equations shows that the additional term $f \delta \lambda$ is always zero because $f = 0$. Hence the arbitrary variation of λ does not affect the variational principle. We therefore notice that the momenta p_i can be regarded as Lagrangian multipliers associated with the constraint

$\dot{q} - \partial H/\partial p_i = 0$. The Lagrangian can therefore be written as

$$L = L + \sum p_i \left(\dot{q}_i - \frac{\partial H}{\partial p_i} \right)$$

$$= \sum p_i q_i + L - \sum p_i \frac{\partial H}{\partial p_i}$$

$$= \sum p_i \dot{q}_i - H \tag{2.68}$$

so that the Legendre transformation can be regarded as the consequence of introducing the Lagrangian multipliers and the equation of constraint. The constraint is purely a matter of the defining equation for the momenta and the mathematical structure of the Legendre transformation. However, we know that physically the Lagrangian multipliers furnish the reaction forces of the system. It is therefore not surprising that the Legendre transformation from velocities to momenta leaves the essential physical content of the variational principle unchanged. It should be noted particularly that each co-ordinate q_i is associated with a particular momentum p_i. The momentum p_i is therefore often called the conjugate momentum of the co-ordinate q_i.

Finally it is important to notice that the new form of the action integral is linear in the velocities. If we regard the momenta as additional co-ordinates, then the 'kinetic energy' is given by the term $\sum p_i \dot{q}_i$ which is particularly simple in structure.

2.5. Duality and energy

By means of the Legendre transformation of equation (2.62) Hamilton's variational principle of stationary action has been transformed so as to depend on the arbitrary variations of the q_i and p_i, the co-ordinates and momenta, instead of the arbitrary variations of the q_i and $\dot{q}_i$, the co-ordinates and velocities. Alternatively, the Legendre transformation can be derived via the method of Lagrange's multipliers where the relationship between the velocities and momenta is introduced as a constraint on the Lagrangian energy function L. The Legendre transformation is obtained by using the momenta as Lagrange's multipliers.

The Euler–Lagrange equations of motion (equation (2.17)) can now be written in terms of the co-ordinates and momenta

$$\dot{p}_i = -\frac{\partial H}{\partial q_i} \tag{2.65}$$

coupled with the set of equations defining the relationship between the velocities and the momenta

$$\dot{q}_i = \frac{\partial H}{\partial p_i}. \tag{2.63}$$

The negative sign in the equations of motion (2.65) is due to the reversed effective forces of d'Alembert's principle. On the other hand, the Legendre transformation is entirely symmetrical and the mathematical choice of the active and passive variables is arbitrary. In the Hamiltonian function H $(q_1, \ldots, q_n; p_1, \ldots, p_n; t)$ the role of the co-ordinates and momenta could be interchanged. The variational principles will be unchanged if both sets of equations (2.63) and (2.65) are unchanged. An interchange between the role of the q_i and p_i must therefore be accompanied by an interchange between $\dot{q}_i$ and $-\dot{p}_i$. The role of each set will have been interchanged but the problem is defined by both sets together.

We can define a new Lagrangian function L' $(p_1, \ldots, p_n; \dot{p}_1, \ldots, \dot{p}_n; t)$ by means of the Legendre transformation

$$L' = \sum_{i=1}^{n} (-q_i \dot{p}_i) - H. \tag{2.69}$$

This Lagrangian is called the *dual* of L $(q_i, \dot{q}_i, t)$. The relationship between the two can be obtained by combining equations (2.62) and (2.69) to give

$$L' = -\sum (q_i \dot{p}_i + p_i \dot{q}_i) + L. \tag{2.70}$$

It will be seen that the action

$$A = \int_{t_1}^{t_2} L \, dt = \int_{t_1}^{t_2} L' \, dt + \left[\sum p_i q_i \right]_{t_1}^{t_2}. \tag{2.71}$$

Hamilton's principle $\delta A = 0$ is therefore unaffected by the substitution of L' for L because the boundary term does not contribute to the variation. The invariance of the two sets of differential equations under the transformation of the co-ordinates and momenta guarantees the variational principle. It will also be seen that the change of signs in the substitution of the $\dot{p}_i$ for the $-\dot{q}_i$ is necessary to render the difference between L and L' integrable. The two differential operators d/dt and $-d/dt$ associated with the transformation of the p_i to the q_i are said to be *adjoint*. This is a property which will become very important when we consider electromagnetic problems later in this book.

So far in this discussion we have treated the transformation of the co-ordinates and momenta as rather a formal mathematical matter. We now briefly enquire into the physical meaning of such transformations. We saw in § 2.2 that in conservative systems, in which L does not depend explicitly on the time t, the total energy of the system defined by equation (2.46) is constant. However, this energy is what we have now defined as the Hamiltonian function

$$H = \sum_{i=1}^{n} \frac{\partial L}{\partial \dot{q}_i} \dot{q}_i - L. \tag{2.72}$$

Hence for conservative systems H is constant and gives the total energy of the system. The constancy of H can be derived in a simple manner by noting that, since H like L does not depend explicitly on t,

$$\frac{\mathrm{d}H}{\mathrm{d}t} = \sum_{i=1}^{n} \left(\frac{\partial H}{\partial q_i} \dot{q}_i + \frac{\partial H}{\partial p_i} \dot{p}_i \right)$$

$$= \sum_{i=1}^{n} (-\dot{p}_i \dot{q}_i + \dot{q}_i \dot{p}_i) = 0. \tag{2.73}$$

If further $L = T - U$ and if T is a quadratic form of the velocities and U is independent of the velocities, we have shown in equation (2.39) that the total energy is given by $T + U$ so that

$$H = T + U. \tag{2.74}$$

The symmetry of the Hamiltonian H in terms of the co-ordinates and momenta is reflected by the fact that H is given by the sum of the energies which is not affected by an interchange between the kinetic momenta and static co-ordinates.

Such a transformation does, however, affect the Lagrangian. Since we have $L = T - U$, the transformation which interchanges the role of the potential and kinetic energies affects the behaviour of the Lagrangian L and its dual L'.

If U is a quadratic form of the co-ordinates

$$U = \tfrac{1}{2} \sum_{i,k=1}^{n} b_{ik} q_i q_k \tag{2.75}$$

and if T is independent of the co-ordinates

$$\dot{p}_i = \frac{\partial L}{\partial q_i} = -\frac{\partial U}{\partial q_i} = -\tfrac{1}{2} \sum_{k=1}^{n} b_{ik} q_k \tag{2.76}$$

Then

$$-\sum_{i=1}^{n} q_i \dot{p}_i = 2U \tag{2.77}$$

and

$$L' = U - T. \tag{2.78}$$

The first variation is $\delta L' = 0$ and is not affected by the change of sign, but the second variation $\delta^2 L'$ is affected because the principle of virtual work given by $\delta \omega \leqslant 0$ for stable equilibrium gives $\delta(T - U) \leqslant 0$ by equation (2.18). This means that

$$\delta^2 L \leqslant 0 \leqslant \delta^2 L'. \tag{2.79}$$

This property of the Lagrangian and its dual makes it possible to obtain upper and lower bounds for the energy of a system near equilibrium. The sign of the forces in d'Alembert's principle embodies a profound physical distinction between potential and kinetic energy. This distinction cannot be removed by a purely mathematical transformation.

2.6. Energy and co-energy

In our examination of mechanical systems we have seen how the integration of the virtual work with respect to time leads to Hamilton's principle of stationary action which is based on the Lagrangian energy L expressed as a function of the co-ordinates q_i and the velocities $\dot{q}_i$. In non-conservative systems L also depends explicitly on the time t, but in this section we shall restrict the discussion to conservative systems.

In § 2.4 we introduced the momenta p_i and showed that the Lagrangian could be transformed into a Hamiltonian energy H expressed as a function of the co-ordinates q_i and the momenta p_i. In § 2.5 we briefly discussed a further transformation which introduced the dual Lagrangian energy L' expressed as a function of the momenta p_i and the forces $\dot{p}_i$.

The Lagrangian L consists of two types of energy, one connected with a system of forces derived from a scalar potential function and the other connected with acceleration of masses. These two types of energy were defined as potential and kinetic energy respectively. In many examples of systems the potential energy is independent of the velocities so that we can write

$$L = T(\dot{q}_i) - U(q_i). \tag{2.80}$$

Similarly

$$H = T(p_i) + U(q_i) \tag{2.81}$$

and

$$L' = U(\dot{p}_i) - T(p_i). \tag{2.82}$$

We proved in § 2.3 that if T is a quadratic form of the velocities as shown in equation (2.47)

$$T(\dot{q}_i) = T(p_i). \tag{2.83}$$

Similarly we proved in § 2.5 that if U is a quadratic form of the co-ordinates as in equation (2.75)

$$U(q_i) = U(\dot{p}_i). \tag{2.84}$$

Hence T and U are invariant under these transformations. However, this invariance depends on the relationships of equations (2.47) and (2.75) and in particular on the fact that the coefficients a_{ik} and b_{ik} are constants. Suppose that we have a single potential energy element. Then equation (2.75) reduces to

$$U = \tfrac{1}{2}bq^2 \tag{2.85}$$

and

$$\dot{p} = \frac{\partial U}{\partial q} = bq. \tag{2.86}$$

The constancy of b therefore implies that the force is proportional to the displacement co-ordinate. Hence the element must have a linear characteristic. If it is non-linear, equation (2.84) does not hold and we must distinguish between the two forms of U. It is convenient to define the potential energy U as $U(q_i)$ and to call $U(\dot{p}_i)$ the potential co-energy U^*. Now for the potential energy we have

$$F_i = -\frac{\partial U}{\partial q_i} \tag{2.3}$$

which in terms of the momenta is

$$\dot{p}_i = \frac{\partial U}{\partial q_i}. \tag{2.87}$$

For the potential co-energy there has been a transformation of q_i to p_i so that

$$q_i = \frac{\partial U^*}{\partial \dot{p}_i}. \tag{2.88}$$

These relationships are illustrated in Fig. 2.1. Formally we have a pleasing

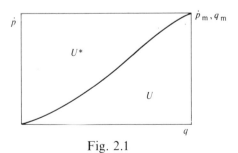

Fig. 2.1

symmetry between equations (2.87) and (2.88), but the physical meaning of
the co-energy is suspect. Whereas the potential energy U is the work done
by the force

$$U = -\int F \, dq \qquad (2.89)$$

we have for the co-energy

$$U^* = \int q \, dF. \qquad (2.90)$$

This is not a mechanical quantity although it has the dimensions of energy.
The co-energy can more instructively be written

$$U^* = |\dot{p}_m q_m| - U. \qquad (2.91)$$

The same argument applies to the kinetic energy. We define the kinetic
energy $T = T(\dot{q})$ and the kinetic co-energy $T^* = T(p)$. From the Euler–
Lagrange equations (2.17) we have

$$F_i = \frac{d}{dt} \frac{\partial T}{\partial \dot{q}} \qquad (2.92)$$

whence

$$p_i = \frac{\partial T}{\partial \dot{q}_i}. \qquad (2.93)$$

From equation (2.63) we obtain

$$\dot{q}_i = \frac{\partial H}{\partial p_i} = \frac{\partial T^*}{\partial p_i}. \qquad (2.94)$$

These relations are illustrated in Fig. 2.2. Since in mechanical systems the
relationship between p and $\dot{q}$ is linear, the magnitudes of T and T^* are

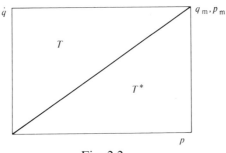

Fig. 2.2

equal. It will be noted that there is a difference between the relative positions of T and T^* and U and U^* in Figs. 2.1 and 2.2. This is reflected in the negative sign $L = T - U$ which arises from d'Alembert's principle. Some writers prefer to write $T(\dot{q})$ as T^* so as to interchange the energy and co-energy. This imposes a formal symmetry, but at the expense of losing the essential difference between potential and kinetic systems. Perhaps the wisest thing is to regard the co-energy as introduced by the transformation

$$T^* = [p_m \dot{q}_m] - T. \tag{2.95}$$

The reader will have noticed that the two figures and equations (2.91) and (2.95) show a simple form of the Legendre transformation by which the energy functions have been transformed.

Finally we note that the dual Lagrangian L' could also be called the co-Lagrangian. The inversion of the potential and kinetic energies from

$$L = T - U$$

to

$$L' = U^* - T^*$$

arises from the fact that we have transformed $\dot{q}$ to p and q to $-\dot{p}$, thus inverting the co-ordinates and velocities. Although this does not affect the equilibrium condition $\delta L = 0 = \delta L'$, it does affect the stability of the equilibrium. If the real system described by L is stable, then the pseudo co-energy system is unstable as shown in equation (2.79).

Appendix

We seek to prove that the necessary and sufficient conditions for the variational principle

$$\delta A = \delta \int_{t_1}^{t_2} L(q_i, \dot{q}_i, t)\, \mathrm{d}t = 0 \tag{2.96}$$

subject to the boundary conditions that the q_is are given at t_1 and t_2 are given by

$$\frac{\mathrm{d}}{\mathrm{d}t} \frac{\partial L}{\partial \dot{q}_i} - \frac{\partial L}{\partial q_i} = 0 \qquad (i = 1, 2, \ldots, n). \tag{2.17}$$

We have

$$\delta L(q_i, \dot{q}_i, t) = L(q_i + \varepsilon \phi_i, \dot{q}_i + \varepsilon \dot{\phi}_i, t) \\ - L(q_i, \dot{q}_i, t) \tag{2.97}$$

where ϕ_i is an arbitrary change in q_i and ε is a quantity which tends to zero. Taking the first term of a Taylor series we have

$$\delta L = \varepsilon \left[\frac{\partial L}{\partial q_i} \phi_i + \frac{\partial L}{\partial \dot{q}_i} \dot{\phi}_i \right]. \tag{2.98}$$

For a stationary value

$$\delta \int_{t_1}^{t_2} L \, dt = \int_{t_1}^{t_2} \delta L \, dt$$

$$= \varepsilon \int_{t_1}^{t_2} \left(\frac{\partial L}{\partial q_i} \phi_i + \frac{\partial L}{\partial \dot{q}_i} \dot{\phi}_i \right) dt = 0. \tag{2.99}$$

The rate of change of the integral must itself be zero, whence

$$\frac{1}{\varepsilon} \delta \int_{t_1}^{t_2} L \, dt = \int_{t_1}^{t_2} \left(\frac{\partial L}{\partial q_i} \phi_i + \frac{\partial L}{\partial \dot{q}_i} \dot{\phi}_i \right) dt = 0. \tag{2.100}$$

This expression can be integrated by parts to give

$$\int_{t_1}^{t_2} \left(\frac{\partial L}{\partial q_i} - \frac{d}{dt} \frac{\partial L}{\partial \dot{q}_i} \right) \phi_i \, dt + \left[\frac{\partial L}{\partial \dot{q}_i} \phi_i \right]_{t_1}^{t_2} = 0 \tag{2.101}$$

but $\phi_i = 0$ at t_1 and t_2, so that

$$\int_{t_1}^{t_2} \left(\frac{\partial L}{\partial q_i} - \frac{d}{dt} \frac{\partial L}{\partial \dot{q}_i} \right) \phi_i \, dt = 0. \tag{2.102}$$

Also ϕ_i is an arbitrary variation of q_i, whence it follows that

$$\frac{\partial L}{\partial q_i} - \frac{d}{dt} \frac{\partial L}{\partial \dot{q}_i} = 0 \tag{2.103}$$

is a necessary condition. Clearly it is also a sufficient condition. Thus we have proved the statement of equation (2.17).

We now seek to prove equation (2.31). From the m constraints

$$f_j(q_1, \ldots, q_n, t) = 0 \qquad (j = 1, 2, \ldots, m) \tag{2.104}$$

we can produce the m variational statements

$$\delta f_j = \frac{\partial f_j}{\partial q_i} \delta q_i = 0. \tag{2.105}$$

These equations hold at any time t. We multiply each equation by a Lagrangian multiplier λ_j and these multipliers will be functions of t. We write

$$\delta \int_{t_1}^{t_2} L \, dt - \int_{t_1}^{t_2} \sum_j \lambda_j \, \delta f_j \, dt = 0. \tag{2.106}$$

The second term is zero by equation (2.105) and can therefore be added to the variation of the first integral. Because there are m constraints we have only $n - m$ free variations of the q_i. The $n - m \, \delta q$ variations can be eliminated by choosing Lagrangian multipliers which cause their coefficients to vanish. The remaining independent δq must then have coefficients which also vanish. Thus the distinction between the $n - m$ free variations and the m constrained variations is purely formal and we can state

$$\frac{\partial L}{\partial q_i} - \frac{d}{dt}\frac{\partial L}{\partial \dot{q}_i} - \lambda_1 \frac{\partial f_1}{\partial q_i} - \lambda_2 \frac{\partial f_2}{\partial q_i} - \ldots - \lambda_m \frac{\partial f_m}{\partial q_i} = 0 \tag{2.31}$$

which is equivalent to replacing L by

$$L' = L - \lambda_1 f_1 - \ldots - \lambda_m f_m. \tag{2.32}$$

Further reading

Most advanced books on mechanics include a section on variational methods.

The variational principles of mechanics, by C. Lanczos, University of Toronto Press, Toronto (1974), is outstanding for its lucidity and enthusiasm.

Dynamics of mechanical and electromechanical systems, by S. H. Crandall *et al.*, McGraw-Hill, New York (1968), is helpful in combining mechanical and electrical systems.

3. Variational Principles in Electromagnetism

3.1. Maxwell's analysis

We have seen in the last chapter that mechanical interactions can be analysed in terms of generalized co-ordinates, velocities, and momenta. In discussing the mechanical behaviour of systems we can solve the complete problem of the motion of such a system using a number of co-ordinates equal to the number of degrees of freedom, which is likely to be much smaller than the number of the parts of the system. All the internal actions and reactions are eliminated at the outset by basing the analysis on the principle of virtual work, which allows us to examine the equilibrium under arbitrary variations of the co-ordinates.

The principle of virtual work, which was originally developed for static systems, can by d'Alembert's remarkable principle of reversing the inertia forces be extended to dynamics. The usefulness of the principle can be even further increased if the virtual work can be integrated. This is possible for static systems if the applied forces can be derived from a work, or potential energy, function. The forces of inertia cannot be derived in this way, but Hamilton overcame the difficulty by integrating the virtual work with respect to time. This leads us to the definition of kinetic energy and to Hamilton's principle of stationary action.

In this chapter we shall examine how electromagnetic interactions can be incorporated in this same scheme. In the historical review of Chapter 1 we have already noted many relationships between electricity and magnetism and ordinary mechanics. There is no doubt that Gilbert regarded the forces between magnets as mechanical forces and in his book he is at great pains to discount 'occult' explanations. Action at a distance was to him a great difficulty, but this difficulty concerned the manner in which the force acted rather than the nature of the force. Newton in his gravitational theory similarly found no difficulty in identifying force between distant masses with force applied to objects in contact. His idea of the mass particles as sources of the force suggested to Priestley the notion of an electric fluid, the droplets of which obeyed the inverse square law of force. There was therefore no difficulty in embodying electric charge, and equally magnetic pole strength, in mechanical systems.

The interaction of electricity and magnetism, however, presented formidable problems. Oersted described his experimental findings in a very obscure terminology. He refers to the 'conflict of electricity which acts on particles of magnetic matter'. Ampere showed that a clearer explanation could be obtained by considering the forces between current elements, but there were strong physical objections to thinking of the fictitious current elements as fundamental particles. It was Faraday who first brought current electricity into the realm of ordinary mechanics by noting that this type of electricity has a property like mechanical momentum or inertia, which also has the remarkable characteristic of distribution in space. Finally Maxwell, taking up Faraday's ideas, came to the conclusion that the energy of current systems was kinetic energy of exactly the same kind as mechanical kinetic energy but that this energy was a property of the system of currents rather than of each particle. Thus kinetic energy was a mutual phenomenon just like potential energy. However, whereas mechanical systems needed mechanical connections to couple the energy of the particles together, electricity acted through space without the need of a coupling material, except perhaps the aether.

It was not surprising that Maxwell turned to Lagrangian mechanics since his kinetic energy was essentially a system parameter. He defined the total kinetic energy as

$$T = T_m + T_e + T_{em} \tag{3.1}$$

where T_m is the mechanical kinetic energy, T_e the electrokinetic energy and T_{em} an energy depending on the interaction of mechanical velocity and current. Maxwell was unable to detect the very small T_{em} effect experimentally and therefore concluded that T_m and T_e could be treated independently so that each current was associated with an extra degree of freedom of the system. He took T_e as a quadratic form of the currents

$$T_e = \tfrac{1}{2} \sum_{i,k=1}^{n} L_{ik} I_i I_k \tag{3.2}$$

where the inductance coefficients L_{ik} correspond to moments and products of inertia of the system of currents.

The momenta could be derived from

$$p_i = \frac{\partial T}{\partial \dot{q}_i} = \frac{\partial T}{\partial I_i} = \sum_{k=1}^{n} L_{ik} I_k. \tag{3.3}$$

Hence the electrokinetic momenta are equal to the magnetic flux linkages Φ_i.

We therefore have a new system of generalized co-ordinates Q_i, velocities $\dot{Q}_i = I_i$, and momenta Φ_i. The dual system which interchanges the co-ordinates and the momenta has Φ_i for the co-odinates, $-\dot{\Phi}_i = V_i$ for the velocities, and Q_i for the momenta. We notice that the voltages V_i are the induced voltages corresponding to a potential energy of the system $-\Sigma V_i Q_i$.

This classification of variables represents a remarkable extension of Lagrangian mechanics. In introducing the electric charges as additional co-ordinates, we now regard as co-ordinates any physical parameters which fit into the Lagrangian scheme and these parameters need not have a geometrical significance. Similarly the generalized velocities need not be actual velocities; they merely have the nature of a kinetic phenomenon. The generalized forces and momenta also have a very much enlarged meaning. This means that the dimensions of these various quantities are changed, but nevertheless the entire behaviour of the system can be described in terms of variational mechanics.

A particularly useful feature of this scheme is that the same equations furnish both mechanical and electrical quantities. For example, the inductance coefficients of equation (3.2) are functions of the geometrical co-ordinates. We can therefore obtain the mechanical forces arising from the position of the currents

$$F_i = \frac{\partial T_e}{\partial q_i} \tag{3.4}$$

as well as the electromotive forces.

Associated with the kinetic energy is the potential energy

$$U_e = \tfrac{1}{2} \sum_{i,k=1}^{n} \frac{1}{C_{ik}} Q_i Q_k \tag{3.5}$$

where C_{ik} are the capacitance coefficients. It is instructive to examine U_e and T_e dimensionally. The C_{ik} vary as the length and the permittivity, which is the product of the relative permittivity which we have taken as independent of the charges and the dimensional constant ε_0. The length is one of the co-ordinates and so we can say that U_e is a function of the electrical and mechanical co-ordinates and the reciprocal of ε_0. Similarly T_e is a function of the electrical velocities, mechanical co-ordinates, and the dimensional constant μ_0. The ratio of T_e to U_e therefore depends on $\mu_0 \varepsilon_0 = 1/c^2$ and from this there follow some important practical results.

The kinetic energy T_e is clearly very much smaller than the potential energy U_e unless the time variation is extremely rapid. Hence the kinetic energy or inductance associated with a set of capacitors can generally be neglected. Similarly the kinetic energy of a current system requires the motion of very large amounts of charge in order to have an appreciable value. Since the currents generally flow in conductors which are electrically neutral, we notice that the positive charges of the stationary lattice have the effect of largely cancelling the potential energy of the moving negative charges. Hence the inductance effect can be treated independently of the potential energy. The cancellation of the potential energies is not complete because there must generally be a very small surface charge to drive the current through the conductor, but these surface charges are not the generalized co-ordinates corresponding to the currents. They are the difference between those co-ordinates and the stationary charges. Hence we can treat inductances as separate objects in the same way as capacitance. However, it is important to remember that this is only an approximation if we are dealing with a dynamic system in which the currents vary. For equilibrium the principle of virtual work demands that the acceleration forces are everywhere balanced by the applied forces. Potential as well as kinetic energy must be distributed throughout the entire system. The reason for the apparent separation of potential and kinetic energy stems from the orders of magnitude involved. Whereas the 'velocity' measured in amperes gives reasonable values of kinetic energy, the corresponding 'co-ordinate' of coulombs gives fantastic values of potential energy, for example the energy of two charges of 1 C each at a distance of 1 m is 9×10^9 J. From the point of view of calculation it is a great help to be able to separate capacitance and inductance effects, but from the point of view of dynamics it is important to remember the close connection between these two types of energy.

Maxwell's scheme of charge as co-ordinate, current as velocity, and flux linkage as momentum deals with 'lumped' circuit parameters and therefore with instantaneous interactions in each element. A full time-varying treatment requires the examination of small regions. In order to do this Maxwell examined the flux linkage with a small 'search coil' and the force with a small 'probe' of charge in an electromagnetic field. This led him to an alternative description of a Lagrangian momentum $\mathbf{A}$, defined by curl $\mathbf{A}$ = $\mathbf{B}$, which he called the electrokinetic momentum at a point. The generalized co-ordinate is still the electric charge Q, the velocity is the current $I = \dot{Q}$ and the momentum is $\mathbf{A}$. The force $-\dot{p} = -\dot{A}$ and this is balanced by the potential energy gradient $-\nabla V$, so that for equilibrium $\nabla V = -\dot{\mathbf{A}}$. It will be noticed that the choice of $\mathbf{A}$ instead of Φ as the momentum involves the replacement of the electromotive force acting around a current loop by the electric field strength acting at a point. Similarly the potential difference V is replaced by its gradient at a point. Dimensionally $\mathbf{A}$ is mechanical momentum per unit charge, but it is important to note that the momentum under consideration here is not the momentum of the test charge itself, since in any case this charge is stationary in its own frame of reference, but is the momentum due to the motion of the other charges which cause the magnetic field. We also note that the electric field strength has the dimensions of mechanical force per unit charge.

Maxwell did not develop the dual system in which the flux linkages are the co-ordinates and the charges the momenta. We notice that, since flux has the dimensions of magnetic pole strength, the dual system is based on the idea of poles or dipoles as being the fundamental particles. The kinetic energy of electric current has been exchanged for the potential energy of magnetic poles and the potential energy of electric charges has been exchanged for the kinetic energy of magnetic currents. The reader may feel that Maxwell was wise in disregarding the somewhat artificial notions introduced by the duality. Like the idea of co-energy these concepts seem to be more trouble than they are worth.

Nevertheless this would be a hasty judgement. It is a fact that the calculation of potential energy is very much simpler than the calculation of kinetic energy because the latter is associated with velocity which is a vector quantity whereas the potential energy depends only on the scalar co-ordinates. The problems of static magnetic fields are therefore generally calculated by the use of a magnetic scalar potential and this implies that the sources of the field are static poles. Maxwell himself used this device, although not in the context of Lagrangian methods.

Maxwell based his choice of electric charge and current as co-ordinate and velocity on the observation that like charges repel and thus reduce their mutual energy, whereas like currents attract and increase the energy of their system. This led him to identify charges as stationary objects having potential energy and currents as implying motion and kinetic behaviour. However, he did not explore the possibility of replacing the electrical sources by equivalent magnetic ones. The dualism draws our attention to the fact that the type of energy is not inherent in the field as such but depends on the physical models we employ for the sources of the field. The possibility of using either electrical or magnetic sources often simplifies calculation and is especially helpful in the method of dual-bounded solutions which we discuss in Chapter 5 and later chapters of this book.

Finally we note that the freedom we have in electromagnetism in interchanging the sources and the types of energy is physically well based. In the last chapter we drew attention to the difference between equations (2.89) and (2.90). Force multiplied by change of co-ordinate is work in the thermodynamic sense, but co-ordinate multiplied by change of force is a quantity involving heat energy. In ordinary mechanics there is a physical difference between co-ordinates and momenta which goes back to d'Alembert's principle. However, in electromagnetism we have a freedom of choice for the sources of the field which enables us to choose a priori which are to be the stationary and which the moving quantities. Even where we are forced to deal with known sources such as known charge and current distributions, the system energy can be treated as if we had an equivalent set of magnetic sources.

3.2. Electrical networks

We have seen that Maxwell's identification of magnetic and kinetic energy and of electric and potential energy enables us to define the circuit elements inductance and capacitance. To complete the description of networks we need to deal with resistance. Ohm's law does not easily fit into the Lagrangian scheme because the electromotive force cannot be derived from either potential or kinetic energy. This is not really surprising because the ohmic process converts energy from electromagnetic to heat energy. We should therefore need thermodynamic terms in order to describe the system in terms of energy. It has already been noted that any electromagnetic system needs mechanical constraints. By itself it is an incomplete subsystem. Now we see the additional need for thermodynamics for a more general description.

Nevertheless we can avoid this complication by reverting to the principle of virtual work in its original differential form. The virtual work of the dissipative ohmic forces can then be accepted as a set of additional terms which cannot be integrated. If the dissipative forces are F_i we can write the Euler–Lagrange equations as

$$F_i = \frac{\mathrm{d}}{\mathrm{d}t} \frac{\partial L}{\partial \dot{q}_i} - \frac{\partial L}{\partial q_i} \tag{3.6}$$

whence

$$RI_i = \frac{\mathrm{d}\Phi_i}{\mathrm{d}t} + V_i. \tag{3.7}$$

If we write the electromotive force E as

$$E = -\frac{\mathrm{d}\Phi}{\mathrm{d}t}$$

we obtain the circuit equations

$$V_i = E_i + R_i I_i. \tag{3.8}$$

Thus this equation is the equation of equilibrium of a system containing potential and kinetic energy and dissipative ohmic forces. We notice that this equation implies equilibrium throughout the system and that the energies and forces are distributed throughout the system. However, we have also noticed that the enormous magnitude of the potential energy compared with the kinetic energy of the same variables, together with the existence of electrically neutral conductors, enables us to segregate the potential and kinetic energies. Similarly we can segregate the dissipative energies. Although resistive current flow requires additional forces for equilibrium, these forces may be associated with amounts of energy small in comparison with the dissipation. Steady electric currents flowing in resistors need a surface electric charge to drive them, but the capacitative effect of these charges is negligible. For time-varying currents the relative magnitudes of the inductive and resistive energies depend on the material and geometrical arrangement, but since the current involved in both processes is the same the actual system can be replaced by a series arrangement of a 'pure' resistance and a 'pure' inductance.

The introduction of the dissipative ohmic forces makes it impossible to use Hamilton's principle $\delta A = 0$ which is based on the variation of the kinetic and potential energies. We can overcome the difficulty by using the principle of virtual work in its differential form which is equivalent to the use of the equations of motion, the Euler–Lagrange equations, rather than the Lagrangian energy itself. However, this is not the only way because it is possible to construct a new variational principle in the following manner.

We consider a network with dissipative elements. In order to achieve an energy balance we introduce sources which exactly provide the ohmic losses. It is convenient, but not essential, to envisage two types of source: ideal voltage generators operating at constant voltage e and varying current, and ideal current generators operating at constant current i and varying voltage. We now require that the energy balance shall be correct not only for the motion as a whole but at each instant of the motion. Thus we arrange for the power to be balanced, the generators supplying exactly the power which is being dissipated, so that the power of the system is zero throughout. Consider a network of l loops which has loop currents I_s, inductances L_{sk}, resistances R_{sk}, capacitances C_{sk} and voltage generators e_k. The Kirchhoff voltage law can be written

$$\sum_{s=1}^{l} \left(R_{sk} I_s + L_{sk} \frac{dI_s}{dt} + \frac{Q_s}{C_{sk}} \right) = e_k \tag{3.9}$$

for $k = 1, 2, \ldots, l$.

We can write this as a principle of virtual power:

$$\sum_{s=1}^{l} \left(R_{sk} I_s \, \delta I_k + L_{sk} \frac{dI_s}{dt} \delta I_k + \frac{Q_s}{C_{sk}} \delta I_k \right) - e_k \, \delta I_k = 0. \tag{3.10}$$

Now put the kinetic energy

$$T = \tfrac{1}{2} \sum_{s,k=1}^{l} L_{sk} I_s I_k \tag{3.11}$$

and the potential energy

$$U = \tfrac{1}{2} \sum_{s,k=1}^{l} \frac{Q_s Q_k}{C_{sk}} \tag{3.12}$$

and put

$$G = \tfrac{1}{2} \sum_{s,k=1}^{l} R_{sk} I_s I_k, \tag{3.13}$$

where G is called the 'content' and is equal to half the instantaneous rate of energy dissipation in the resistive elements. Then the virtual power can be written

$$\frac{\partial G}{\partial I_k} \delta I_k + \frac{\mathrm{d}}{\mathrm{d}t} \frac{\partial T}{\partial I_k} \delta I_k + \frac{\partial U}{\partial Q_k} \delta I_k - e_k \, \delta I_k = 0. \tag{3.14}$$

However,

$$\frac{\mathrm{d}}{\mathrm{d}t} \left(\frac{\partial T}{\partial I_k} \right) = \frac{\partial}{\partial I_k} \left(\frac{\mathrm{d}T}{\mathrm{d}t} \right) \tag{3.15}$$

and

$$\frac{\partial U}{\partial Q_k} = \frac{\partial}{\partial I_k} \left(\frac{\mathrm{d}U}{\mathrm{d}t} \right) \tag{3.16}$$

so that equation (3.11) becomes

$$\frac{\partial G}{\partial I_k} \delta I_k + \frac{\partial}{\partial I_k} \left(\frac{\mathrm{d}T}{\mathrm{d}t} + \frac{\mathrm{d}U}{\mathrm{d}t} \right) \delta I_k - e_k \, \delta I_k = 0 \tag{3.17}$$

and we can write the variational principle

$$\delta \left| 2G + \frac{\mathrm{d}}{\mathrm{d}t} (T + U) - P \right| = 0 \tag{3.18}$$

where P is the power supplied by the generators. Thus the Kirchhoff voltage law can be obtained from the variational statement of equation (3.18). The second variation is positive because only G is quadratic in the currents and its coefficients are essentially positive. Hence the power dissipated has a minimum stationary value under variations of the loop currents. It will be noted that the potential energy is related to the charges and the kinetic energy to the currents.

The dual power principle can be derived in terms of the node potentials and the power supplied by current generators.

The Kirchhoff current law can be written

$$\sum_{s=1}^{n} \left(\frac{E_s}{R_{sk}} + \frac{\Phi_s}{L_{sk}} + C_{sk} \frac{\mathrm{d}E_s}{\mathrm{d}t} \right) = i_k. \tag{3.19}$$

It can be transformed into a principle of virtual power by writing

$$\sum_{s=1}^{n} \left(\frac{E_s}{R_{sk}} \delta E_k + \frac{\Phi_s}{L_{sk}} \delta E_k + C_{sk} \frac{\mathrm{d}E_s}{\mathrm{d}t} \delta E_k \right) - i_k \, \delta e_k = 0. \tag{3.20}$$

Now we put

$$T' = \tfrac{1}{2} \sum_{s_1 k = 1}^{n} C_{sk} E_s E_k \tag{3.21}$$

$$U' = \tfrac{1}{2} \sum_{s_1 k = 1}^{n} \frac{\Phi_s \Phi_k}{L_{sk}} \tag{3.22}$$

$$G' = \tfrac{1}{2} \sum_{s_1 k = 1}^{n} \frac{E_s E_k}{R_{sk}} \tag{3.23}$$

and we can write the variational principle

$$\delta \left\{ 2G' + \frac{\mathrm{d}}{\mathrm{d}t} (T' + U') - P \right\} = 0. \tag{3.24}$$

The dissipated power is again a minimum. It will be noted that the variational principle now relates potential energy to the magnetic flux and kinetic energy to the voltages. It is important to notice that in the statements of equations (3.18) and (3.24) we find the sum of the potential and kinetic energies, whereas Hamilton's principle uses the difference. In our present discussion we are dealing with the variation of power rather than energy. Such a restricted variational principle has been forced upon us because we have energy dissipation in the resistors. This means that the energy balance must be correct at each instant of time and it is not possible to find a single variational statement to cover the behaviour throughout time. As a result of these investigations the variational principle can also be stated in the very general form as

$$\delta \sum_{\alpha = 1}^{n} I_\alpha V_\alpha = 0 \tag{3.25}$$

where the elements of the network are denoted by the αs and the sign of the generated power is taken as opposite to that of the power in the passive elements. The two Kirchhoff laws can then be obtained as the equations of equilibrium of the system. It should be noted that the voltages and currents are varied independently of each other so that the currents I_α need not be the actual currents corresponding to the voltages V_α, and *vice versa*. All that is required is that the currents satisfy the current law and the voltages the voltage law. It should be noted that this form of the variational principle depends on the segregation of the energies so that each element of the system has two terminals between which there is a single potential difference.

Equation (3.25) is known as Tellegen's theorem and is often stated without the variational δ. In such a statement it merely implies the obvious truth that the instantaneous power of an isolated network is zero. The real strength of the theorem arises from its variational nature and our derivation has endeavoured to bring out the connection of the theorem with Lagrangian mechanics. This relationship is particularly useful if the systems under consideration have both electromagnetic and mechanical components or if it is desired to analyse mechanical systems in terms of equivalent electromagnetic systems or networks. We shall examine these matters shortly, but before doing so we must mention a very useful feature of Tellegen's theorem, namely that the voltages and currents in the elements do not have to be related linearly. This is of course a direct result of the status of the theorem as a variational principle which holds for arbitrary independent variations of the voltages and currents. The relationship between voltage and current does not enter into the variation. We have already discussed this in § 2.6 where we examined the effect of interchanging the co-ordinates and momenta. There we showed that we could deal with such a transformation by exchanging the energy and co-energy (equations (2.87)–(2.88)). The same is true for Tellegen's theorem. Suppose we consider this with reference to the power in the resistances, and suppose the relationship between voltage and currents is as shown in Fig. 3.1, then the content G is shown as the area $\int_0^I V \, \mathrm{d}I$ and the co-content G' is the area $\int_0^V I \, \mathrm{d}V$. Thus we use G for problems which obtain the Vs by varying the Is and G' for the dual problem of obtaining the Is by varying the Vs. Since it is clear that we can vary the function only when it is expressed in terms of the varying parameter, there is no difficulty and it seems hardly worthwhile to introduce the additional term 'co-content'. For linear systems the two quantities are in any case the same. It should, however, be noted explicitly that the variational method can deal with non-linear elements and this is one of its great advantages. The non-linearity is restricted, however, by the fact that the elements are passive. This means that the slope of the curve between voltage and current must not become negative as shown in Fig. 3.2. Such a curve would make it impossible to specify a unique value for $\int I \, \mathrm{d}v$

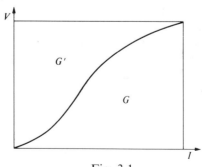

Fig. 3.1

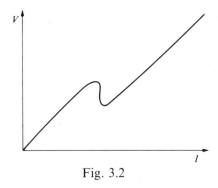

Fig. 3.2

and the variational principle cannot be applied. Physically such an element has a region of negative resistance, which would of course alter the equilibrium conditions.

We now return to the consideration of the equivalence of electrical and mechanical energy in networks. The natural choice is to follow Maxwell and equate potential energy with capacitance and kinetic energy with inductance, thus regarding charge as a stationary co-ordinate and magnetic flux as a momentum. Unfortunately this choice is topologically unsatisfactory. Geometrical co-ordinates involve distance and are thus connected with two points. In a similar way potential differences involve two points, so that both are called 'across' variables. Similarly mechanical momentum is associated with mass which, since mass is conserved in a flow system, makes it a 'through' quantity like electric charge. In this nomenclature we have the following classification of variables shown in Table 3.1. To maintain the same topological structure for the mechanical and electric networks we need therefore to consider springs and inductances together as possessing potential energy and similarly moving masses and capacitances as possessing kinetic energy. The alternative, dynamically more satisfactory, comparison of springs with capacitances and moving masses with inductances results in an interchange of series and parallel connections. The choice of flux as a stationary co-ordinate and charge as momentum implies that magnetic pole strength is the fundamental quantity and deals

TABLE 3.1

Variables	Mechanical	Electrical
Across state variable	Co-ordinate q	Magnetic flux Φ
Across rate variable	Velocity $\dot{q}$	Voltage $-\dot{\Phi}$
Through state variable	Momentum p	Charge Q
Through rate variable	Force $-\dot{p}$	Current $I = \dot{Q}$

with electromagnetic relationships as the dynamics of electric poles. In network problems the conceptual difficulties are lessened because there is no dynamical interaction between the types of energy, each element having only one type. Thus the only thing that needs to be specified is a pair of quantities, one being of the 'through' variety and the other being an 'across' variable.

The description of electromagnetic systems in terms of networks of single-energy elements is extremely powerful in modelling the topology, but it necessarily imposes restrictions on the type of physical process which can be dealt with. Tellegen's theorem of invariant power implies that the currents and voltages are determined at every instant and moreover that the currents are continuous and that the voltages are potential differences rather than electromotive forces. The continuity of current implies instantaneous interaction throughout the system as well as in each element, so that the model cannot deal with wave motion which depends on a finite propagation time. The replacement of electromotive forces by potential difference makes the voltages independent of the path along which they act. This obliterates the distinction between kinetic and potential energy. Also the elements of such networks are one dimensional and have no internal structure. The resistances, inductances, and capacitances have lost their primary meaning as energy parameters and have become transfer functions expressing the relationships between the 'through' and the 'across' variables.

Finally we note that, since the physical separation of potential and kinetic energy and dissipation is never complete, an actual network of capacitors, inductors, and resistors will behave differently from the idealized capacitance, inductance, and resistance network. In such a network the energies will interact and as a result the interaction will become time dependent and will exhibit a 'wave' type of behaviour. Tellegen's theorem will then be only approximately true.

3.3. Distributed systems

We now leave the discussion of systems consisting of idealized elements and examine the mechanics of distributed systems. In other words we pass from circuit theory to field theory. In § 3.1 we discussed Maxwell's use of the 'electrokinetic momentum at a point' **A**, which enabled him to deal with the mechanics of a point charge. Now we want to discuss the mechanics of systems of distributed charge density and current density. This discussion is to be in terms of the Lagrangian method of considering the variation of the energy of the system and, because the system is distributed in space, we shall have to consider the energy as the space integral of an energy density. Because of this additional complication we shall proceed cautiously and shall consider static systems before dealing with time-varying systems.

3.3.1. Electrostatic systems

The field quantities in electrostatics are the charge density ρ which is associated with the electric flux density $\mathbf{D}$ by the relationship

$$\nabla \cdot \mathbf{D} = \rho \tag{3.26}$$

and the electric field strength $\mathbf{E}$ which is associated with the scalar potential ϕ by the relationship

$$-\nabla\phi = \mathbf{E}. \tag{3.27}$$

The energy volume density is $\frac{1}{2}\mathbf{E} \cdot \mathbf{D}$ or $\frac{1}{2}\phi\rho$. We wish to fit these quantities into the Lagrangian scheme of co-ordinates and momenta. Since equations (3.26) and (3.27) contain space derivatives rather than time derivatives, we must make the necessary transposition in the application of the method. Let us first choose ϕ as a generalized co-ordinate and $\mathbf{D}$ as a generalized momentum. This is a reasonable choice since ϕ is a scalar function of position and $\mathbf{D}$ is a vector associated with the electric flux. We now have $\mathbf{E}$ as the equivalent of a velocity where the space operator $-\nabla$ has replaced the time operator $\mathrm{d}/\mathrm{d}t$. Similarly the charge density ρ has taken the place of the force $-\dot{p}$ and the operator $\nabla \cdot$ has been interchanged with $-\mathrm{d}/\mathrm{d}t$. The proposed scheme is therefore as follows:

Co-ordinate	Velocity	Momentum	Force
q	$\dot{q}$	p	$-\dot{p}$
ϕ	$\mathbf{E}$	$\mathbf{D}$	ρ

The object of our investigation is the energy of the electrostatic system under consideration. This is a scalar quantity and can be expressed as a definite integral of an energy density expressed in terms of the field parameters, the integration having to be performed throughout the space occupied by the system. The energy is therefore a functional of the field parameters and corresponds to the action A in Lagrangian mechanics. The change from an integration in a fixed interval of time to an integration in a fixed volume affects the way in which the parameters have to be varied and it also affects the way in which the system can be specified.

There are two ways in which we can specify the system, either by stating the assigned charge density $\rho = \bar{\rho}$ or the assigned potential distribution $\phi = \bar{\phi}$ of the charge density. Suppose we choose the charge density; then the Lagrangian variation can be carried out in ϕ and $\mathbf{E}$ where $\mathbf{D}$ is related to $\mathbf{E}$ and will vary with $\mathbf{E}$. The stationary value for the variation will then be given by equation (3.26). Alternatively we can vary $\mathbf{D}$, ρ and $\mathbf{E}$ as functions of $\mathbf{D}$ and keep ϕ at its assigned value $\bar{\phi}$. This will furnish the dual Lagrangian variation.

Consider first an electrostatic system having an assigned charge density. We have

$$\mathbf{V} \cdot \mathbf{D} = \bar{\rho}. \tag{3.28}$$

Let us treat this as a variational principle akin to the principle of virtual work. We have

$$\langle (\bar{\rho} - \mathbf{V} \cdot \mathbf{D}), \delta\phi \rangle = 0 \tag{3.29}$$

where the $\langle \; \rangle$ notation represents the volume integration of the scalar product.

We now seek to integrate the virtual work so as to express it as the variation of a scalar energy quantity. To do so we make use of the vector identity

$$\nabla \cdot (u\mathbf{A}) = \mathbf{A} \cdot \mathbf{V}u + u\mathbf{V} \cdot \mathbf{A} \tag{3.30}$$

which gives us

$$\langle \mathbf{D}, \mathbf{V}\delta\phi \rangle - [D_n, \delta\phi] + \langle \bar{\rho}, \delta\phi \rangle = 0 \tag{3.31}$$

where [ ] represents a surface integration. We shall deal with this surface term in detail in the next chapter in our discussion of Green's theorem, but here we shall ignore it noting that it could be zero if $\mathbf{D} = 0$ at the surface, i.e. if the bounding surface is outside the field. Hence

$$-\langle \mathbf{D}, \delta\mathbf{E} \rangle + \langle \bar{\rho}, \delta\phi \rangle = 0 \tag{3.32}$$

and

$$-\tfrac{1}{2}\varepsilon\langle E^2 \rangle + \langle \bar{\rho}, \phi \rangle = \text{constant.} \tag{3.33}$$

The first term contains the electrostatic field energy and the second is twice the assembly work of an isolated system of the charges $\bar{\rho}$. However, the assembly work is equal to the field energy, so that the energy of the system is finally given by

$$U = \langle \bar{\rho}, \phi \rangle - \tfrac{1}{2}\varepsilon\langle E^2 \rangle \tag{3.34}$$

and the variational principle is

$$\delta U = 0 \tag{3.35}$$

where U is a functional of ϕ and $\mathbf{E}$, and of $\mathbf{D} = \varepsilon\mathbf{E}$.

The dual specification is

$$\langle (\nabla\phi + \mathbf{E}), \delta\mathbf{D} \rangle = 0 \tag{3.36}$$

$$\langle -\phi, \mathbf{V} \cdot \delta\mathbf{D} \rangle + [\phi, \delta D_n] + \langle \mathbf{E}, \delta\mathbf{D} \rangle = 0 \tag{3.37}$$

and ignoring the boundary term

$$\langle -\bar{\phi}, \delta\rho \rangle + \langle \mathbf{E}, \delta\mathbf{D} \rangle = 0 \tag{3.38}$$

whence

$$\langle -\bar{\phi}, \rho \rangle + \frac{1}{2\varepsilon} \langle D^2 \rangle = \text{constant} \tag{3.39}$$

so that finally

$$\delta U' = 0 \tag{3.40}$$

where

$$U' = \langle \bar{\phi}, \rho \rangle - \frac{1}{2\varepsilon} \langle D^2 \rangle. \tag{3.41}$$

We notice that the second variations $\delta^2 U$ and $\delta^2 U'$ both depend on the second terms of the expressions and are therefore both negative. Hence U and U' are both maxima for the specified variations.

The reader may agree with all this but may have doubts about what has been achieved. The usefulness of the method will become very apparent in Chapter 5, but we can already give some indication of why the method has value.

Suppose we again focus attention on the energy U (or U') of the system, which we seek to calculate. If we start with equations (3.26) and (3.27), these can be combined into Poisson's equation

$$-\nabla^2\phi = \rho/\varepsilon. \tag{3.42}$$

This equation can be integrated subject to given boundary conditions. The integration process furnishes the values of the potential ϕ throughout the region occupied by the system, but we do not need all this information if we merely seek the single scalar energy. We can obtain the energy by calculating the electric field $\mathbf{E}$ from the potential distribution and thence the energy density $\frac{1}{2}\mathbf{E} \cdot \mathbf{D}$. By integrating this we finally arrive at the energy. The integration process will average the energy density and will therefore destroy much of the information laboriously gathered by the calculation of ϕ. Physically ϕ gives the energy per unit charge and the calculation of ϕ corresponds to the probing of the system by means of a small point charge. However, this point charge has nothing to do with the system itself nor with its energy. Hence the process is unnecessary as well as cumbersome.

The variational method starts at the other end by calculating a system energy which has a stationary value subject to an equilibrium condition. In this programme equations (3.26) and (3.27) are subsidiary conditions needed to obtain the value of the energy of the system in a state of equilibrium. Of course we shall not have calculated the field by this process, but if we do not need to know the field we shall have saved a great deal of labour and expense.

It is important to notice that the success of the variational scheme depends on the possibility of integrating the virtual work. This depended on the vector identity of equation (3.30) and on the relationship between the vector operators $-\nabla$ and $\nabla \cdot$. These two operators are said to be adjoint. In Lagrangian mechanics we met the same problem in having to integrate the virtual work of the inertia forces. There we noted that

$$\frac{\mathrm{d}}{\mathrm{d}t}(qp) = p\dot{q} + q\dot{p} \tag{3.43}$$

which enabled us to put $-q(\mathrm{d}p/\mathrm{d}t)$ for $+p(\mathrm{d}q/\mathrm{d}t)$ in the integration process. The operators $+\mathrm{d}/\mathrm{d}t$ and $-\mathrm{d}/\mathrm{d}t$ are adjoint and similarly the operators $-\nabla$ and $\nabla \cdot$ are adjoint. We could therefore write the Lagrangian parameters more generally as

$$q, \quad Tq, \quad p, \quad T^{\mathrm{a}}p$$

where T and T^{a} are adjoint operators defined by

$$\langle p, Tq \rangle = \langle q, T^{\mathrm{a}}p \rangle + \text{boundary term.} \tag{3.44}$$

The Euler–Lagrange equations can then be written as

$$T^{\mathrm{a}} \frac{\partial L}{\partial (Tq)} + \frac{\partial L}{\partial q} = 0 \tag{3.45}$$

$$T \frac{\partial L'}{\partial (T^{\mathrm{a}}p)} + \frac{\partial L'}{\partial p} = 0. \tag{3.46}$$

In our treatment so far we have assumed that there is a linear relationship between $\mathbf{D}$ and $\mathbf{E}$. As discussed in § 3.2 this is an unnecessary restriction. All that needs to be done is to distinguish between the energy

$$\left\langle \int_0^E \mathbf{D} \cdot \delta\mathbf{E} \right\rangle$$

in equation (3.29) and the 'co-energy'

$$\left\langle \int_0^D \mathbf{E} \cdot \delta\mathbf{D} \right\rangle$$

in equation (3.38). The only restriction is that the integrals inside the brackets must have unique values; thus the relationship between D and E

must be of the form shown in Fig. 3.1 and not as shown in Fig. 3.2. The reader should satisfy himself that the Euler–Lagrange equations, which in this instance are given by equations (3.45) and (3.46), are not affected by the non-linearity. This property of the variational method is one of its great advantages. The method which starts with the differential equations runs into the considerable difficulty that in Poisson's equation (3.42) the permittivity ε will depend on the field if there is a non-linear relation between **E** and **D**. This means that we have to write

$$-\mathbf{V} \cdot (\varepsilon \mathbf{V} \phi) = \rho \qquad (3.47)$$

and this equation is considerably more difficult to integrate.

It is helpful to enquire into the physical meaning of the dual Lagrangian formulations. In ordinary mechanics there is a clear distinction between co-ordinates and momenta. Is there an equivalent distinction in the formulation of electrostatic problems? We have so far chosen ϕ as a co-ordinate and **D** as momentum. This fits the idea of flux and also corresponds to the geometrical ideas of 'across' and 'through' quantities which we discussed in § 3.2. However, it also implies that **E** is a velocity for which there seems to be no good reason. Suppose instead we consider **D** to be the co-ordinate, ρ the velocity, ϕ the momentum and **E** the force. This fits the idea that the electric field is the force on a unit charge, an idea which is of great importance in electromagnetism. On the other hand, it conflicts with geometrical ideas. Unfortunately we cannot have it both ways.

The dual system of **E** as force has as its basis the idea of a point charge. What is the basis of the system having **E** as a velocity? The answer to this question can be seen by considering the equivalent magnetic field problem. The magnetic field strength **H** can arise either as force per unit pole or as m.m.f. gradient in A/m, i.e. line density of electric current. Thus **E** could be expressed as line density of magnetic current. Now magnetic current would be the flow of pole strength and naturally **E** would be associated with velocity. Hence the original assumption of ϕ as co-ordinate and **D** as momentum has as its physical basis the idea of magnetic pole strength, whereas the dual **D** as co-ordinate and ϕ as momentum has electric charge as its basis. The energy of the electrostatic field can equally well be described as the kinetic energy of steady magnetic currents or as the potential energy of stationary electric charges. The former choice has geometrical advantages and the latter has the physical advantage of using charge rather than the somewhat dubious magnetic current.

Mathematically the electric field has divergence and curl sources. The divergence sources furnish the potential energy and the curl sources the kinetic energy. The Lagrangian formulation invites us to choose either type of energy. This can be done because the energy can be expressed purely in terms of **E** and **D** without explicit use of the sources of the field. It should be noted that this possibility exists only for the energy and not for the field by itself because the measurement of the field involves a probe which is also a source. The choice of probe involves the choice of co-ordinate and momentum. The Lagrangian method is firmly based on the idea of system energy, whereas the concept of field often lacks physical basis because the necessity for interaction with a probe is easily forgotten.

Before we leave this topic there is one further matter to be investigated. The choice of co-ordinate and momentum defines the variational process but it does not define the specification of the problem. For example, a capacitor problem could be specified in terms of either charge or potential without determining the choice of which of these is to be the co-ordinate and which the momentum.

In equation (3.34) we specify $\bar{\rho}$ and vary ϕ and **E**, and in equation (3.41) we specify $\bar{\phi}$ and vary **D** and ρ. There are two other possibilities. We can specify $\bar{\rho}$ on the capacitor but vary **D** but not ρ between the plates, or we can specify $\bar{\phi}$ on the plates but allow ϕ and **E** to vary between the plates. We shall discuss these and other processes more fully in the next chapter. Here we confine ourselves to a formal treatment.

Consider equation (3.37) but leave ϕ unspecified. We have

$$-\langle \phi, \delta\rho \rangle + [\phi, \delta D_n] + \langle \mathbf{E}, \delta\mathbf{D} \rangle = 0. \tag{3.48}$$

The first two terms are zero since $\delta\rho$ is zero in the volume and $\delta\mathbf{D}$ is zero on the plates. Hence the variational statement is

$$\delta\left(\frac{1}{2\varepsilon}\langle D^2 \rangle\right) = 0 \tag{3.49}$$

and we note that the second variation is positive. Similarly equation (3.31) becomes

$$-\langle \mathbf{D}, \delta\mathbf{E} \rangle - [D_n, \delta\phi] + \langle \rho, \delta\phi \rangle = 0. \tag{3.50}$$

On the plates $\delta\phi = 0$ if ϕ is specified there. Also ρ in the volume is zero. Hence only the first term remains. The variational statement becomes

$$\delta\left(\frac{\varepsilon}{2}\langle E^2 \rangle\right) = 0 \tag{3.51}$$

and the second variation of the energy is positive. We therefore have four possibilities for the variational statements. Two of these include the sources explicitly and give a negative value for the second variation. They therefore specify a maximum for the energy. The other statements give the energy without explicit reference to the sources and specify a minimum energy. We shall make good use of these possibilities in calculating approximate values for the energy in Chapter 5.

3.3.2. Magnetostatic systems

If there is no electric current inside the volume occupied by the system we have

$$\mathbf{V} \times \mathbf{H} = 0 \tag{3.52}$$

and hence we can use a magnetic scalar potential ϕ_m defined by

$$-\mathbf{V}\phi_m = \mathbf{H}.$$

The Lagrangian variables are then

q	Tq	p_m	$T^a p$
ϕ_m	$\mathbf{H}$	$\mathbf{B}$	ρ_m

where the magnetic pole density ρ_m is given by

$$\mathbf{V} \cdot \mathbf{B} = \rho_m. \tag{3.53}$$

This conflicts with the usual statement

$$\mathbf{V} \cdot \mathbf{B} = \dot{0} \tag{3.54}$$

which is used if the magnetic field is taken as arising from current sources. The use of the scalar potential ϕ_m is an artifice and the pole density is to be understood in terms of Ampère's equivalence between current loops and magnetic dipoles. The artifice is particularly useful in calculating systems including iron.

The variational principle is

$$\delta U_m = 0 \tag{3.55}$$

where

$$U_m = \langle \phi_m, \bar{\rho}_m \rangle - \tfrac{1}{2}\mu \langle H^2 \rangle \tag{3.56}$$

with its dual

$$\delta U_m' = 0$$

where

$$U_m' = \langle \phi_m, \rho_m \rangle - \frac{1}{2\mu} \langle B^2 \rangle. \tag{3.57}$$

Alternatively we can start with the current sources of density $\mathbf{J}$ given by

$$\mathbf{V} \times \mathbf{H} = \mathbf{J}. \tag{3.58}$$

Since

$$\mathbf{V} \cdot \mathbf{B} = 0 \tag{3.54}$$

we can define a vector potential $\mathbf{A}$ by

$$\nabla \times \mathbf{A} = \mathbf{B} \tag{3.59}$$

and

$$\nabla \cdot \mathbf{A} = 0. \tag{3.60}$$

We then have the Lagrangian scheme

q	Tq	p	$T^a p$
$\mathbf{A}$	$\mathbf{B}$	$\mathbf{H}$	$\mathbf{J}$

where $T = T^a = \nabla \times$, these operators being not merely adjoint but 'self-adjoint' as can be seen from the vector identity

$$\nabla \cdot (\mathbf{A} \times \mathbf{J}) = \mathbf{J} \cdot (\nabla \times \mathbf{A}) - \mathbf{A} \cdot (\nabla \times \mathbf{J}). \tag{3.61}$$

The variational principle becomes

$$\delta T_m = 0 \tag{3.62}$$

where

$$T_m = \langle \mathbf{A}, \bar{\mathbf{J}} \rangle - \tfrac{1}{2}\mu \langle B^2 \rangle \tag{3.63}$$

and the dual

$$\delta T_m' = 0 \tag{3.64}$$

where

$$T_m' = \langle \bar{\mathbf{A}}, \mathbf{J} \rangle - \tfrac{1}{2}\mu \langle H^2 \rangle. \tag{3.65}$$

We have used the kinetic energy T_m rather than the potential energy U_m because the sources of T_m are currents whereas the sources of U_m are poles.

As shown in § 3.3.1 two further principles can be derived which exclude explicit reference to the sources. They are

$$\delta \left(\frac{1}{2\mu} \langle B^2 \rangle \right) = 0 \tag{3.66}$$

and

$$\delta \left(\frac{\mu}{2} \langle H^2 \rangle \right) = 0. \tag{3.67}$$

3.3.3. Resistive systems with steady current flow

We have already discussed the problem of the dissipation of energy by ohmic resistance in § 3.2 when we dealt with networks having lumped parameters. There we saw that for equilibrium the energy had to be supplied to the system at the same rate as it was dissipated. Hence we had to deal with a power balance rather than an energy balance. The same is true for distributed resistance in a resistive system.

Ohm's law can be written

$$\sigma \mathbf{E} = \mathbf{J} \tag{3.68}$$

where σ is the conductivity.

Since we are here dealing with steady conditions which do not vary with time

$$\mathbf{V} \times \mathbf{E} = 0 \tag{3.69}$$

and

$$-\mathbf{V}\phi = \mathbf{E}. \tag{3.27}$$

The electric field is therefore an electrostatic field and its sources are stationary electric charges. The equation of continuity of electric current flow is

$$\mathbf{V} \cdot \mathbf{J} + \frac{\partial \rho}{\partial t} = 0. \tag{3.70}$$

Hence for steady current flow .

$$\mathbf{V} \cdot \mathbf{J} = 0 \tag{3.71}$$

and there can be no charge density in the volume of the system. This means that the sources of the electric field must be outside the region of current flow or on its surface. These charges have capacitance energy and we notice that the separation between resistance and capacitance cannot be complete. However, the capacitance energy is constant in time and will not further concern our consideration of the power balance. In any case the values of conductivity are such as to make the charges very small. We already considered this in § 3.2 and noticed that it enabled us to separate the various types of energy process in electromagnetism. In our present discussion the power has to be supplied to the system by maintaining the current flow at the surfaces.

Consider the Lagrangian system

q	Tq	p	$T^{\mathrm{a}}p$
ϕ	$\mathbf{E}$	$\mathbf{J}$	0

where $T = -\mathbf{V}$ and $T^{\mathrm{a}} = \mathbf{V} \cdot$ are the adjoint operators. The variational principle is given by

$$\langle \mathbf{V} \cdot \mathbf{J}, \delta\phi \rangle = 0. \tag{3.72}$$

Using the adjoint relationship, we have

$$\langle \mathbf{J}, \delta \mathbf{E} \rangle + [J_n, \delta \phi] = 0 \qquad (3.73)$$

where J_n is the normal component of $\mathbf{J}$ of the surface. This can be integrated, if we use Ohm's law in the volume term and treat $J_n = \bar{J}_n$ as the assigned surface current. Hence

$$\tfrac{1}{2}\sigma \langle E^2 \rangle + [J_n, \phi] = \text{constant.} \qquad (3.74)$$

However,

$$[J_n, \phi] = -\langle \mathbf{J}, \mathbf{E} \rangle = -\sigma \langle E^2 \rangle \qquad (3.75)$$

so that the power, or content, of the system is

$$G = -[\bar{J}_n, \phi] - \tfrac{1}{2}\sigma \langle E^2 \rangle \qquad (3.76)$$

and

$$\delta G = 0. \qquad (3.77)$$

The dual system is given by

$$\delta G' = 0 \qquad (3.78)$$

where

$$G' = -[J_n, \phi] - \frac{1}{2\sigma} \langle J^2 \rangle. \qquad (3.79)$$

Both $\delta^2 G < 0$ and $\delta^2 G' < 0$, so that G and G' are maxima for those variational processes. This result should be compared with equations (3.18) and (3.24) where the power dissipation was a minimum in processes which varied either the loop currents or the node potentials of a network. The reason for this difference is that the networks had no external sources. G and G' are then defined entirely by the second terms of equations (3.76) and (3.79).

3.3.4. Time-varying systems having inductance and resistance

We now face the problem of having to deal with systems which vary in time as well as being distributed in space, whereas so far we have dealt with lumped-parameter time-varying systems or distributed static systems. Let us restrict the investigation initially to deal with quasi-static systems for which the displacement current is negligible. The equations are as follows:

$$\nabla \times \mathbf{H} = \mathbf{J} \qquad (3.58)$$

$$\nabla \times \mathbf{E} = -\frac{\partial \mathbf{B}}{\partial t} \qquad (3.80)$$

$$\sigma \mathbf{E} = \mathbf{J} \qquad (3.68)$$

$$\mathbf{B} = \mu \mathbf{H}. \qquad (3.81)$$

Equation (3.58) implies that

$$\mathbf{V} \cdot \mathbf{J} = 0. \tag{3.71}$$

This implies that there is no volume charge density ρ since in general the equation of continuity gives

$$\mathbf{V} \cdot \mathbf{J} + \frac{\partial \rho}{\partial t} = 0 \tag{3.70}$$

and the absence of charge density gives

$$\mathbf{V} \cdot \mathbf{D} = 0 \tag{3.82}$$

which is consistent with

$$\mathbf{D} = \varepsilon \mathbf{E} \tag{3.83}$$

and Ohm's law (equation (3.68)).

The four parameters describing the system of equations (3.58), (3.80), (3.68) and (3.81) are $\mathbf{H}$, $\mathbf{J}$, $\mathbf{E}$ and $\mathbf{B}$ and we wish to embody these in the Lagrangian scheme. However, we notice that the system is not adjoint. If we combine the four equations into two

$$\mathbf{V} \times \mathbf{H} = \mathbf{J} \tag{3.58}$$

and

$$\mathbf{V} \times \mathbf{J} = -\sigma \mu \frac{\partial \mathbf{H}}{\partial t}. \tag{3.84}$$

The operator $\mathbf{V} \times$ is self-adjoint, but there is nothing to correspond to the operator $-\partial/\partial t$. Let us consider this from a physical point of view. The system has magnetic energy varying with time and it also has dissipation varying with time. We have seen previously that a dissipative system requires sources of energy before we can discuss its equilibrium. Since by Ohm's law the dissipation is stated in terms of the current flow, and hence in terms of the rate of energy dissipation, we have to consider a principle of virtual power rather than virtual work so that the system is in equilibrium at every instant. In § 3.3.3 where we considered steady current flow the power could be put into the system at its boundaries, but with a time-varying system in which the energy diffuses at a certain speed this is no longer possible if equilibrium is to be maintained. In such a system the power must be supplied in exactly the same manner as it is dissipated both in space and time. This means that we must devise an adjoint system which generates power so that the two systems together will be in equilibrium. Such a conceptual system can easily be devised by endowing it with a negative conductivity, and we then have for the adjoint system

$$\mathbf{V} \times \mathbf{H}^* = \mathbf{J}^* \tag{3.85}$$

$$\mathbf{V} \times \mathbf{J}^* = \sigma\mu \frac{\partial \mathbf{H}^*}{\partial t}. \tag{3.86}$$

Equations (3.58) and (3.84) and equations (3.85) and (3.86) can be combined into

$$\mathbf{V} \times \mathbf{V} \times \mathbf{H} = -\sigma\mu \frac{\partial \mathbf{H}}{\partial t} \tag{3.87}$$

and

$$\mathbf{V} \times \mathbf{V} \times \mathbf{H}^* = +\sigma\mu \frac{\partial \mathbf{H}^*}{\partial t}. \tag{3.88}$$

These equations could also be written in terms of the other field quantities $\mathbf{J}$, $\mathbf{J}^*$ and $\mathbf{E}$, $\mathbf{E}^*$ and $\mathbf{B}$, $\mathbf{B}^*$. Mathematically the difference between equations (3.87) and (3.88) is in the sign of the operator $\partial/\partial t$. Because of this it is convenient to think of the adjoint system as having a negative time sequence. This also fits the energy conservation law because the adjoint system supplies the energy which the ordinary system dissipates. Instead of dealing with negative conductivity we prefer to deal with a negative time sequence. If the time variation is harmonic all the field quantities in the real and adjoint systems then have the same magnitude.

A principle of virtual power is given by

$$\left\langle \frac{1}{\sigma} \mathbf{V} \times \mathbf{V} \times \mathbf{H} + \mu \frac{\partial \mathbf{H}}{\partial t}, \delta\mathbf{H}^* \right\rangle = 0. \tag{3.89}$$

Using the vector identity of equation (3.61) we can write this as

$$\left\langle \frac{1}{\sigma} \mathbf{V} \times \mathbf{H}, \mathbf{V} \times \delta\mathbf{H}^* \right\rangle$$
$$+ \left[\left(\frac{1}{\sigma} \mathbf{V} \times \mathbf{H} \right) \times \delta\mathbf{H}^*, \hat{\mathbf{n}} \right] + \left\langle \mu \frac{\partial \mathbf{H}}{\partial t}, \delta\mathbf{H}^* \right\rangle = 0 \tag{3.90}$$

whence

$$\langle \mathbf{E}, \delta\mathbf{J}^* \rangle + [\mathbf{E} \times \delta\mathbf{H}^*, \hat{\mathbf{n}}] + \left\langle \mu \frac{\partial \mathbf{H}}{\partial t}, \delta\mathbf{H}^* \right\rangle = 0. \tag{3.91}$$

Before we can integrate this we must give further attention to the significance of the adjoint system. This system had to be introduced in order to provide the instantaneous energy balance. The products of the quantities of the original system and the adjoint system are therefore designed to give a power which is invariant in time and this is the significance of the variational statement of equation (3.91). This can be seen most easily if we consider harmonic variation in time. The original field quantities will then vary as exp $(j\omega t)$ and the adjoint quantities as exp $(-j\omega t)$, which will make their products invariant with time. If we introduce this harmonic time variation into equation (3.91) we have

$$\frac{1}{\sigma} \langle \mathbf{J}, \delta \mathbf{J}^* \rangle + [\mathbf{E} \times \delta \mathbf{H}^*, \hat{\mathbf{n}}] + j\omega\mu \langle \mathbf{H}, \delta \mathbf{H}^* \rangle = 0 \qquad (3.92)$$

where the adjoint quantities are the complex conjugates of the original quantities. If now the surface $\mathbf{E}$ is prescribed as $\bar{\mathbf{E}}$, we can integrate and derive the variational statement

$$\delta \left\{ -\frac{1}{2\sigma} \langle J^2 \rangle - [\bar{\mathbf{E}} \times \mathbf{H}^*, \hat{\mathbf{n}}\rangle - \frac{j\omega\mu}{2} \langle H^2 \rangle \right\} = 0 \qquad (3.93)$$

where $\langle J^2 \rangle$ and $\langle H^2 \rangle$ are integrals of the squares of the magnitude of $\mathbf{J}$ and $\mathbf{H}$ and the sign of the surface term has been chosen to give the input to the system.

A dual principle can be derived by varying $\mathbf{E}$ and $\mathbf{B}$:

$$\left\langle \frac{1}{\mu} \nabla \times \nabla \times \mathbf{E}^* - \sigma \frac{\partial \mathbf{E}^*}{\partial t}, \delta \mathbf{E} \right\rangle = 0 \qquad (3.94)$$

$$\left\langle \frac{1}{\mu} \nabla \times \nabla \times \mathbf{E}^* - j\omega\sigma \mathbf{E}^*, \delta \mathbf{E} \right\rangle = 0 \qquad (3.95)$$

$$\frac{1}{\mu} \langle \nabla \times \mathbf{E}^*, \nabla \times \delta \mathbf{E} \rangle - j\omega[\delta \mathbf{E} \times \mathbf{H}^*, \hat{\mathbf{n}}] - j\omega\sigma \langle \mathbf{E}, \delta \mathbf{E}^* \rangle = 0 \qquad (3.96)$$

$$\frac{\omega^2}{\mu} \langle \mathbf{B}^*, \delta \mathbf{B} \rangle - j\omega[\delta \mathbf{E} \times \mathbf{H}^*, \hat{\mathbf{n}}] - j\omega\sigma \langle \mathbf{E}, \delta \mathbf{E}^* \rangle = 0. \qquad (3.97)$$

If H is prescribed as $\bar{H}$ on the surface of the system, we can integrate and derive the variational principle

$$\delta \left\{ -\frac{\sigma}{2} \langle E^2 \rangle - [\mathbf{E} \times \bar{\mathbf{H}}^*, \hat{\mathbf{n}}] - \frac{j\omega}{2\mu} \langle B^2 \rangle \right\} = 0. \qquad (3.98)$$

A specification of $\bar{\mathbf{E}}$ on the surface corresponds to an applied voltage and a specification of $\bar{\mathbf{H}}$ to an applied current. We can therefore write the variational principle of equation (3.93) as

$$\frac{V^2}{2} \delta \left\{ \frac{1}{r} + j \frac{1}{x} \right\} = 0 \qquad (3.99)$$

where r and x form a parallel circuit having a common voltage V. The principle of equation (3.98) can be written as

$$\frac{I^2}{2} \, \delta\{R + jX\} = 0 \qquad (3.100)$$

where R and X form a series circuit having a common current I. We shall discuss the question of upper and lower bounds for the circuit parameters in § 5.10.

Two further variational principles can be derived which omit the surface term. They are

$$\delta\left\{\frac{1}{2\sigma} \langle J^2 \rangle + \frac{j\omega\mu}{2} \langle H^2 \rangle\right\} = 0 \qquad (3.101)$$

and

$$\delta\left\{\frac{\sigma}{2} \langle E^2 \rangle - \frac{j\omega}{2\mu} \langle B^2 \rangle\right\} = 0. \qquad (3.102)$$

It is of great interest to note that the introduction of complex conjugate quantities, which is such a familiar feature of power calculations in a.c. theory, is a consequence of the need for an adjoint system without which it is not possible to derive the field equations from the variation of an invariant functional which is a circuit parameter.

3.3.5. Electromagnetic waves

In this section we discuss systems having inductance, capacitance and radiated power. We have

$$\nabla \times \mathbf{H} = \mathbf{J} + \frac{\partial \mathbf{D}}{\partial t} \qquad (3.103)$$

$$\nabla \times \mathbf{E} = -\frac{\partial \mathbf{B}}{\partial t} \qquad (3.80)$$

$$\nabla \cdot \mathbf{B} = 0 \qquad (3.56)$$

$$\nabla \cdot \mathbf{D} = \rho \qquad (3.26)$$

$$\mathbf{D} = \varepsilon \mathbf{E} \qquad (3.83)$$

$$\mathbf{B} = \mu \mathbf{H}. \qquad (3.81)$$

In order to derive a variational principle for the power we consider the virtual power

$$\left\langle \nabla \times \mathbf{H} - \mathbf{J} - \frac{\partial \mathbf{D}}{\partial t}, \delta \mathbf{E}^* \right\rangle = 0 \qquad (3.104)$$

where $\mathbf{E}^*$ is the adjoint electric field. Hence

$$\langle \mathbf{H}, \delta\mathbf{V} \times \mathbf{E}^* \rangle - [\delta\mathbf{E}^* \times \mathbf{H}, \hat{\mathbf{n}}] - \langle \mathbf{J}, \delta\mathbf{E}^* \rangle - \left\langle \frac{\partial \mathbf{D}}{\partial t}, \delta\mathbf{E}^* \right\rangle = 0 \qquad (3.105)$$

and therefore

$$+ \left\langle \mathbf{H}, \delta \frac{\partial \mathbf{B}^*}{\partial t} \right\rangle - \left\langle \frac{\partial \mathbf{D}}{\partial t}, \delta\mathbf{E}^* \right\rangle - [\delta\mathbf{E}^* \times \mathbf{H}, \hat{\mathbf{n}}] - \langle \mathbf{J}, \delta\mathbf{E}^* \rangle = 0.$$
$$(3.106)$$

If $\mathbf{J} = \bar{\mathbf{J}}$ is assigned in the volume and $\mathbf{H} = \bar{\mathbf{H}}$ on the surface of the system and if the time variation is harmonic, we can write the variational statement

$$\delta\left\{ \left(\frac{j\omega}{2\mu} \langle B^2 \rangle - \frac{j\omega\varepsilon}{2} \langle E^2 \rangle \right) - [\mathbf{E}^* \times \bar{\mathbf{H}}, \hat{\mathbf{n}}] - \langle \mathbf{E}^*, \bar{\mathbf{J}} \rangle \right\} = 0. \qquad (3.107)$$

The dual virtual power is given by

$$\left\langle \left(\mathbf{V} \times \mathbf{E} + \frac{\partial \mathbf{B}}{\partial t} \right), \delta\mathbf{H}^* \right\rangle = 0 \qquad (3.108)$$

where $\mathbf{H}^*$ is the adjoint magnetic field. Hence

$$\langle \mathbf{E}, \delta\mathbf{V} \times \mathbf{H}^* \rangle + [\mathbf{E} \times \delta\mathbf{H}^*, \hat{\mathbf{n}}] + \left\langle \frac{\partial \mathbf{B}}{\partial t}, \delta\mathbf{H}^* \right\rangle = 0 \qquad (3.109)$$

and therefore

$$\langle \mathbf{E}, \delta\mathbf{J}^* \rangle + [\mathbf{E} \times \delta\mathbf{H}^*, \hat{\mathbf{n}}] - \left\langle \delta \frac{\partial \mathbf{D}^*}{\partial t}, \mathbf{E} \right\rangle + \left\langle \frac{\partial \mathbf{B}}{\partial t}, \delta\mathbf{H}^* \right\rangle = 0.$$
$$(3.110)$$

If $\mathbf{E} = \bar{\mathbf{E}}$ both in the volume and on the surface of the system and if the time variation is harmonic, we can write the variational statement

$$\delta\left\{ \langle \bar{\mathbf{E}}, \mathbf{J} \rangle + [\mathbf{E} \times \mathbf{H}^*, \hat{\mathbf{n}}] - \frac{j\omega}{2\varepsilon} \langle D^2 \rangle + \frac{j\omega\mu}{2} \langle H^2 \rangle \right\} = 0. \qquad (3.111)$$

Equations (3.107) and (3.111) are statements about the power. To obtain a variational energy principle we make use of the potentials $\mathbf{A}$ and ϕ defined by

$$\mathbf{V} \times \mathbf{A} = \mathbf{B} \qquad (3.59)$$

$$\mathbf{E} = -\frac{\partial \mathbf{A}}{\partial t} - \mathbf{V}\phi. \qquad (3.112)$$

These two equations do not define the potentials uniquely because they are unchanged by the transformations

$$\mathbf{A}' = \mathbf{A} - \mathbf{V}\psi \tag{3.113}$$

$$\phi' = \phi + \frac{\partial \psi}{\partial t}. \tag{3.114}$$

If ψ is chosen to obey the homogeneous wave equation, which means that its sources are outside the region of interest,

$$\nabla^2 \psi - \frac{1}{c^2} \frac{\partial^2 \psi}{\partial t^2} = 0 \tag{3.115}$$

$$\mathbf{V} \cdot (\mathbf{A}' - \mathbf{A}) + \frac{1}{c^2} \frac{\partial}{\partial t} (\phi' - \phi) = 0. \tag{3.116}$$

If therefore the potentials are linked by the Lorentz condition

$$\mathbf{V} \cdot \mathbf{A} + \frac{1}{c^2} \frac{\partial \phi}{\partial t} = 0 \tag{3.117}$$

the transformation of the potentials by equations (3.113) and (3.114) leaves their definition unchanged. Under these conditions the potentials constitute a 'four-vector' which assures the correct relationship of the space and time relationships of relativity theory. This is, however, a digression, and all that we need to note here is that the variations of $\mathbf{A}$ and ϕ are linked by equation (3.117). When we have obtained the variational principle we shall test it to see whether it satisfies the relationship between $\mathbf{A}$ and ϕ.

We now proceed with the establishment of the variational principle. Combining equations (3.59) and (3.112) we can write

$$\begin{bmatrix} \mathbf{B} \\ \mathbf{E} \end{bmatrix} = \begin{bmatrix} \mathbf{V} \times & 0 \\ -\dfrac{\partial}{\partial t} & -\mathbf{V} \end{bmatrix} \begin{bmatrix} \mathbf{A} \\ \phi \end{bmatrix} \tag{3.118}$$

and combining equations (3.103) and (3.26)

$$\begin{bmatrix} \mathbf{J} \\ \rho \end{bmatrix} = \begin{bmatrix} \mathbf{V} \times & -\dfrac{\partial}{\partial t} \\ 0 & \mathbf{V} \cdot \end{bmatrix} \begin{bmatrix} \mathbf{H} \\ \mathbf{D} \end{bmatrix}. \tag{3.119}$$

Examination of the operator matrices shows that these two sets of equations are not adjoint because of the sign of the time differentiation. To ensure adjointness we can modify them and write

$$
\begin{bmatrix} \mathbf{B} \\ -\mathbf{E} \end{bmatrix} = \begin{bmatrix} \mathbf{\nabla} \times \\ \dfrac{\partial}{\partial t} \quad \mathbf{\nabla} \end{bmatrix} \begin{bmatrix} \mathbf{A} \\ \phi \end{bmatrix} \tag{3.120}
$$

$$
\begin{bmatrix} \mathbf{J} \\ -\rho \end{bmatrix} = \begin{bmatrix} \mathbf{\nabla} \times & \dfrac{\partial}{\partial t} \\ & -\mathbf{\nabla} \cdot \end{bmatrix} \begin{bmatrix} \mathbf{H} \\ \mathbf{D} \end{bmatrix}. \tag{3.121}
$$

Consider first a variation of the potentials by means of a principle of virtual energy

$$
\left\langle \left(\mathbf{\nabla} \times \mathbf{H} - \frac{\partial \mathbf{D}}{\partial t} - \mathbf{J} \right), \delta \mathbf{A} \right\rangle - \langle (\mathbf{\nabla} \cdot \mathbf{D} - \rho), \delta \phi \rangle = 0 \tag{3.122}
$$

where the negative sign of the second term has been chosen in order to associate $+\mathbf{J}$ with $-\rho$ as in equation (3.121). We notice that the integration has to be carried out both in space and time. To make this explicit we write

$$
\int \left\{ \left(\mathbf{\nabla} \times \mathbf{H} - \frac{\partial \mathbf{D}}{\partial t} - \mathbf{J} \right) \cdot \delta \mathbf{A} - (\mathbf{\nabla} \cdot \mathbf{D} - \rho) \, \delta \phi \right\} dv \, dt
$$

$$
= \int \left(\mathbf{H} \cdot \delta \mathbf{B} + \mathbf{D} \cdot \delta \frac{\partial \mathbf{A}}{\partial t} - \mathbf{J} \cdot \delta \mathbf{A} + \mathbf{D} \cdot \delta \mathbf{\nabla} \phi + \rho \delta \phi \right) dv \, dt
$$

$$
+ \int (\mathbf{H} \times \delta \mathbf{A}) \cdot \hat{\mathbf{n}} \, ds \, dt - \int D_n \, \delta \phi \, ds \, dt - \int \mathbf{D} \cdot \delta \mathbf{A} \, dv. \tag{3.123}
$$

Using equation (3.112) to link the variations in $\mathbf{A}$ and ϕ and assuming $\mathbf{J} = \bar{\mathbf{J}}$ and $\rho = \bar{\rho}$ in the volume and $\mathbf{H} = \bar{\mathbf{H}}$ and $D_n = \bar{D}_n$ on the surface of the system we obtain

$$
\int \left(\frac{1}{2\mu} B^2 - \frac{\varepsilon}{2} E^2 - \bar{\mathbf{J}} \cdot \mathbf{A} + \bar{\rho} \phi \right) dv \, dt
$$

$$
+ \int |(\bar{\mathbf{H}} \times \mathbf{A}) \cdot \hat{\mathbf{n}} - \bar{D}_n \phi| \, ds \, dt - \int \mathbf{D} \cdot \delta \mathbf{A} \, dv = \text{constant}. \tag{3.124}
$$

To eliminate the last term we impose the usual Lagrangian condition that $\mathbf{A}$ is not varied at the beginning and end of the process as we did in equations (2.13) and (2.14). Hence finally

$$
\delta \left\{ \langle \bar{\mathbf{J}}, \mathbf{A} \rangle - \langle \bar{\rho}, \phi \rangle - \frac{1}{2\mu} \langle B^2 \rangle + \frac{\varepsilon}{2} \langle E^2 \rangle - |\bar{\mathbf{H}} \times \mathbf{A}, \hat{\mathbf{n}}| + [\bar{D}_n, \phi] \right\} = 0 \tag{3.125}
$$

where we have changed the signs in order to give a positive sign to the kinetic energy density $\langle \bar{\mathbf{J}}, \mathbf{A} \rangle$ and a negative sign to the potential energy density $\langle \bar{\rho}, \phi \rangle$.

The dual variational principle can be derived as follows:

$$\langle (\nabla \times \mathbf{A} - \mathbf{B}), \delta \mathbf{H} \rangle + \left\langle \left(\mathbf{E} + \frac{\partial \mathbf{A}}{\partial t} + \nabla \phi \right), \delta \mathbf{D} \right\rangle = 0 \qquad (3.126)$$

$$\int \left(\mathbf{A} \cdot \delta \mathbf{J} + \mathbf{A} \cdot \frac{\partial \delta \mathbf{D}}{\partial t} - \mathbf{B} \cdot \delta \mathbf{H} + \mathbf{E} \cdot \delta \mathbf{D} - \mathbf{A} \cdot \frac{\partial \delta \mathbf{D}}{\partial t} - \phi \delta \rho \right) dv \, dt$$

$$+ \int (\mathbf{A} \times \delta \mathbf{H}) \cdot \hat{\mathbf{n}} \, ds \, dt + \int \phi D_n \, ds \, dt + \int \mathbf{A} \cdot \delta \mathbf{D} \, dv = \text{constant.} \qquad (3.127)$$

Assuming that the assigned sources are given by $\mathbf{A} = \bar{\mathbf{A}}$ and $\phi = \bar{\phi}$ we obtain

$$\int \left(\bar{\mathbf{A}} \cdot \mathbf{J} - \frac{\mu}{2} H^2 + \frac{1}{2\varepsilon} D^2 - \bar{\phi} \rho \right) dv \, dt$$

$$+ \int (\bar{\mathbf{A}} \times \mathbf{H}) \cdot \hat{\mathbf{n}} \, ds \, dt + \int \bar{\phi} D_n \, ds \, dt + \int \mathbf{A} \cdot \delta \mathbf{D} \, dv = \text{constant.} \qquad (3.128)$$

If we impose the condition that $\mathbf{D}$ is not varied at the beginning and end of the process we have

$$\delta \left\{ \langle \bar{\mathbf{A}}, \mathbf{J} \rangle - \langle \bar{\phi}, \rho \rangle - \frac{\mu}{2} \langle H^2 \rangle + \frac{1}{2\varepsilon} \langle D^2 \rangle - [\mathbf{H} \times \bar{\mathbf{A}}, \hat{\mathbf{n}}] + [\phi, D_n] \right\} = 0.$$

$$(3.129)$$

It will be noticed that the two variational statements of equations (3.125) and (3.129) embody the variational principles of electrostatics (equations (3.35) and (3.40)) and of magnetostatics (equations (3.62) and (3.64)).

Finally we apply the test of equations (3.113) and (3.114) to the potentials $\mathbf{A}$ and ϕ. Consider first equation (3.125) for the two sets of potentials $\mathbf{A}, \phi$ and $\mathbf{A}', \phi'$. We then consider the difference of two variational statements given by

$$\delta \left\{ -\langle \bar{\mathbf{J}}, \nabla \psi \rangle - \left\langle \rho, \frac{\partial \psi}{\partial t} \right\rangle + [\bar{\mathbf{H}} \times \nabla \psi, \hat{\mathbf{n}}] + \left[\bar{D}_n, \frac{\partial \psi}{\partial t} \right] \right\}$$

$$= \delta \int \left(-\mathbf{J} \cdot \nabla \psi - \frac{\partial \psi}{\partial t} + \mathbf{J} \cdot \nabla \psi + \nabla \psi \cdot \frac{\partial \mathbf{D}}{\partial t} + \bar{\rho} \frac{\partial \psi}{\partial t} + \mathbf{D} \cdot \nabla \frac{\partial \psi}{\partial t} \right) dv \, dt$$

$$= \delta \int \mathbf{D} \cdot \nabla \psi \, dv = \delta \int \mathbf{D} \cdot (\mathbf{A} - \mathbf{A}') \, dv = 0 \qquad (3.130)$$

since $\mathbf{A} = \mathbf{A}'$ at the beginning and end of the process and $\mathbf{D}$ is not varied. Consider next the variation of equation (3.129). In this $\bar{\mathbf{A}}$ and $\bar{\phi}$ are assigned values which are not varied. We consider the possibility of two different sets $\bar{\mathbf{A}}$, ϕ and $\bar{\mathbf{A}}'$, ϕ'. Then

$$\delta\left\{-\langle \nabla\psi, \mathbf{J}\rangle - \left\langle \frac{\partial\psi}{\partial t}, \rho\right\rangle + [\mathbf{H} \times \nabla\psi, \hat{\mathbf{n}}] + \left[\frac{\partial\psi}{\partial t}, D_n\right]\right\}$$

$$= \delta\int\left(-\nabla\psi\cdot\mathbf{J} - \rho\frac{\partial\psi}{\partial t} + \nabla\psi\cdot\mathbf{J} + \nabla\psi\cdot\frac{\partial\mathbf{D}}{\partial t} + \rho\frac{\partial\psi}{\partial t} + \mathbf{D}\cdot\nabla\frac{\partial\psi}{\partial t}\right) dv\, dt$$

$$= \delta\int\mathbf{D}\cdot\nabla\psi\, dv = 0 \qquad\qquad (3.131)$$

since $\nabla\psi$ is not varied throughout the motion and $\mathbf{D}$ is not varied at the beginning and end of the motion.

Hence the variational principles of the electromagnetic energy are not affected by the transformation of the potentials in accordance with equations (3.113) and (3.114). This is as it should be, since the transformation of the potentials leaves the magnetic and electric fields unchanged and can therefore have no measurable effect.

Once again it is possible to construct a number of other variational principles and in particular the source terms can be excluded. This may be necessary if, as is often the case, the specification of the problem is in terms of one set of sources such as the potentials, but not in terms of the currents and charges, or *vice versa*. The dual principle must then omit the unknown sources, but their energy will be included implicitly in the field terms. This is a very general feature of the method and will be illustrated by examples in Chapter 5.

Further reading

J. C. Maxwell's famous two-volume *Treatise on electricity and magnetism*, Clarendon Press, Oxford (1891) (reprinted by Dover Publications, New York, 1954) is surprisingly easy to read and very rewarding. Our treatment is based on Part 4 of the treatise, particularly Chapters 5 and 6.

An interesting and extensive discussion of Tellegen's theorem is given by P. Penfield, R. Spence and S. Duinker, *Tellegen's theorem and electrical networks*, MIT Press, Cambridge, Mass. (1970).

4. Some General Energy Theorems

In the last chapter we used the principle of virtual work to derive variational statements for various electromagnetic processes and systems. The method we employed was based on selecting one or more of Maxwell's equations and integrating these in a manner which was analogous to that used in Lagrangian mechanics. Although the method was straightforward and had no special mathematical or conceptual difficulty, it may have seemed to be rather formal and analytical. We were manipulating symbols rather than concentrating our attention on the underlying physical processes. Although the manipulation of the electromagnetic quantities is very important in the application of the variational method described in this book, we need to be sure of the underlying physical meaning before we can apply the method with any degree of confidence. In this chapter we intend to concentrate on this meaning by taking up some well-known general theorems and explaining their relevance to the variational method. It is hoped that by this means the reader will be able to relate the method to his existing knowledge of electromagnetism.

4.1. Green's theorem and the definition of electrostatic systems

George Green's essay on *The application of mathematical analysis to the theories of electricity and magnetism* was published privately in 1828 and is perhaps the most important paper in the history of electrostatics except for the original work of Priestley and Cavendish on the inverse-square law. It might have been lost if a copy of it had not been shown to William Thomson, later Lord Kelvin, while he was a student at Cambridge in 1845. Green's work inspired Kelvin and formed the starting point of many of his contributions to electrostatics including the work on the theory of electrostatic images.

From the point of view of this book it is interesting to note that the motive for Green's investigation was to submit the 'phenomena of the equilibrium of electric and magnetic fluids' to mathematical analysis. Green looked to mechanical principles for his inspiration and his essay is a remarkable example of physical insight guiding mathematical skill. It is convenient to approach Green's work by stating it mathematically and then finding the physical content of the mathematics.

Let v be a closed region of space bounded by a surface s and let ϕ and ψ be two scalar functions of position which are continuous and have continuous first and second differentials within v. Apply the divergence theorem to the vector $\phi\nabla\psi$:

$$\int_v \nabla\cdot(\phi\nabla\psi)\,dv = \oint_s (\phi\nabla\psi)\cdot ds \tag{4.1}$$

whence

$$\int_v (\phi\nabla^2\psi + \nabla\phi\cdot\nabla\psi)\,dv = \oint_s \phi\frac{\partial\psi}{\partial n}\,ds. \tag{4.2}$$

Equation (4.2) is often referred to as Green's first identity. Suppose now that ϕ is the electrostatic potential and $\nabla\psi = \mathbf{D}$ the electrostatic flux density. Then $\nabla^2\psi = \rho$ and $\nabla\phi = -\mathbf{E}$, so that equation (4.2) becomes

$$\int_v (\phi\rho - \mathbf{E}\cdot\mathbf{D})\,dv = \oint_s \phi D_n\,ds. \tag{4.3}$$

If we write

$$\tfrac{1}{2}\int_v (\phi\rho - \mathbf{E}\cdot\mathbf{D})\,dv = \tfrac{1}{2}\oint_s \phi D_n\,ds \tag{4.4}$$

we note that the left-hand side is the difference between the assembly work of the charge density ρ and the field energy $\tfrac{1}{2}\mathbf{E}\cdot\mathbf{D}$ and that this difference is equal to a surface term given by the right-hand side. Suppose now that we replace D_n by a surface charge density σ where $D_n = -\sigma$. This terminates the field $\mathbf{D}$ on the charge density as illustrated in Fig. 4.1. Equation (4.4) can be rewritten

$$\tfrac{1}{2}\int_v (\phi\rho)\,dv + \tfrac{1}{2}\oint_s \phi\sigma\,ds - \tfrac{1}{2}\int_v \mathbf{E}\cdot\mathbf{D}\,dv = 0. \tag{4.5}$$

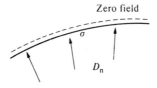

Fig. 4.1

This can be interpreted by saying that the surface s has been replaced by a surface s' outside s and that the surface σ has now been included in the region surrounded by s' (see Fig. 4.2). Thus there is no contribution to the energy from the surface s'. The system of charges ρ and σ is isolated from the rest of space and this enables its internal energy to be specified uniquely. The surface charge σ disconnects the system from the outside world, or alternatively we can regard the effect of the outside world as equivalent to the surface charge $\sigma = -D_n$. Since electric charge acts through empty space,

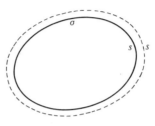

Fig. 4.2

this notional surface charge is not only useful but necessary in order to define the energy of the electrostatic system. If the third term in equation (4.5) is transferred to the other side,

$$\tfrac{1}{2} \int_v \phi\rho \; dv + \tfrac{1}{2} \oint_s \phi\sigma \; ds = \tfrac{1}{2} \int_v \mathbf{E} \cdot \mathbf{D} \; dv. \qquad (4.6)$$

The right-hand side is the field energy of the system and we note that the field energy is equal to the assembly work, and hence to the potential energy, if σ is included. Without the inclusion of σ neither the assembly work nor the field energy are uniquely defined for the system of charges.

The reader will have noticed that there is an alternative way of disconnecting the system from the outside world. The surface potential ϕ can be replaced by a double layer of charge equal to a step of potential. This formulation corresponds to the dual formulation of the last chapter in which the potentials rather than the charges are taken as the assigned sources. We can then write

$$-\tfrac{1}{2} \int_v \varepsilon\phi \; \nabla^2\phi \; dv + \tfrac{1}{2} \oint_s \varepsilon\phi \frac{\partial\phi}{\partial n} \; ds = \tfrac{1}{2} \int_v \mathbf{E} \cdot \mathbf{D} \; dv \qquad (4.7)$$

where ε is the permittivity.

The uniqueness of the energy can be formally proved as follows. Consider a system of charges of density ρ within a volume v and suppose that there are two potential functions ϕ_1 and ϕ_2 such that

$$\nabla^2\phi_1 = \nabla^2\phi_2 = -\rho/\varepsilon \qquad (4.8)$$

within v.

Green's first identity gives for both ϕ_1 and ϕ_2

$$\int_v \{\phi\nabla^2\phi + (\nabla\phi)^2\}\,dv = \oint_s \phi\nabla\phi\cdot ds. \qquad (4.9)$$

Put $\chi = \phi_1 - \phi_2$. Then from equations (4.8) and (4.9)

$$\int_v (\nabla\chi)^2\,dv = \oint_s \chi\nabla\chi\cdot ds. \qquad (4.10)$$

The surface term is zero if ϕ or $\nabla\phi$ are specified on the surface because then $\phi_1 = \phi_2$ or $\nabla\phi_1 = \nabla\phi_2$. Then the volume term must also be zero and this implies that

$$\nabla\chi = 0. \qquad (4.11)$$

Hence

$$\phi_1 - \phi_2 = \text{constant}. \qquad (4.12)$$

If ϕ is specified on the surface $\phi_1 = \phi_2$ and the constant is zero. The field energy is constant with either surface specification. The assembly work depends on the original position of the charges and it is uniquely equal to the field energy if the original position is such that there is no net charge distribution. It will be noticed that the total charge of an isolated system is necessarily zero, so that

$$\int_v \rho\,dv + \oint_s \sigma\,ds = 0. \qquad (4.13)$$

The assembly of the charges thus involves a process of separating positive from negative charge. The more usual argument of bringing charges together from infinity cannot be applied to bounded systems.

4.2. Mutual energy and Green's functions

So far in this chapter we have considered the total energy of a system of charges. Green's theorem also enables us to define the mutual energy of two or more electrostatic systems which are bounded by the same surface.

Consider Green's first identity given by equation (4.2). If we transpose ϕ and ψ we have

$$\int_v (\psi \nabla^2 \phi + \nabla \psi \cdot \nabla \phi)\, dv = \oint_s \psi \frac{\partial \phi}{\partial n}\, ds. \qquad (4.14)$$

If we subtract equation (4.14) from equation (4.2) we obtain

$$\int_v (\phi \nabla^2 \psi - \psi \nabla^2 \phi)\, dv = \oint_s \left(\phi \frac{\partial \psi}{\partial n} - \psi \frac{\partial \phi}{\partial n} \right) ds \qquad (4.15)$$

which is known as Green's second identity or Green's theorem.

If the volume charge density giving rise to the potential ϕ is ρ and its equivalent surface charge density is σ, and if ψ is associated with ρ' and σ', we have

$$\int_v (\phi \rho' - \psi \rho)\, dv = -\oint_s (\phi \sigma' - \psi \sigma)\, ds \qquad (4.16)$$

or

$$\int_v \phi \rho'\, dv + \oint_s \phi \sigma'\, ds = \int_v \psi \rho\, dv + \oint_s \psi \sigma\, ds. \qquad (4.17)$$

This is the statement that the mutual energy of the two systems can be calculated either by inserting the charges ρ', σ' into the field ϕ due to ρ and σ, or by reversing the process and inserting ρ, σ into the field ψ. If the surface is at a large distance from the charges, ϕ and ψ decrease at least as $1/r$ and $\partial \psi / \partial n$ and $\partial \psi / \partial n$ as $1/r^2$ so that the surface integral decays as $1/r$ and can be neglected. We then have

$$\int_v \phi \rho'\, dv = \int_v \psi \rho\, dv \qquad (4.18)$$

where the volume integral includes all the sources ρ and ρ'. In some examples one or other of the integrals may be easier to calculate and this may be a considerable help in finding the mutual capacitance.

Another useful result follows if we regard one of the systems of charges as a means of probing the field of the other charges. Suppose for example that we choose for our probe a small region of high charge density, the total charge being unity. This can be regarded as a 'point' unit charge.

Mathematically this is a Dirac δ function defined by the functional properties

$$\delta(r) = 0 \qquad r \neq 0 \tag{4.19}$$

$$\int \delta(r)\, dv' = 1 \tag{4.20}$$

and

$$\int f(x', y', z')\delta(r)\, dv' = f(x, y, z) \tag{4.21}$$

where the point $r = 0$ is included in the volume of integration in equations (4.20) and (4.21), x, y, z are the co-ordinates of the Dirac function, and f is any arbitrary function of the co-ordinates.

The potential ψ of a small unit charge is given by

$$\psi = \frac{1}{4\pi\varepsilon_0 r}. \tag{4.22}$$

In spherical co-ordinates

$$\nabla^2\psi = \frac{1}{r^2}\frac{\partial}{\partial r}\left(r^2\frac{\partial\psi}{\partial r}\right). \tag{4.23}$$

If $r \neq 0$, $\nabla^2\psi = 0$ as expected because there is no charge density ρ' anywhere except at $r = 0$. If $r = 0$ is included in the volume of integration

$$\int_{v'} \nabla^2\left(\frac{1}{4\pi\varepsilon_0 r}\right) dv' = \frac{1}{4\pi\varepsilon_0}\int_{s'} \nabla\left(\frac{1}{r}\right) ds' = -\frac{1}{\varepsilon_0}. \tag{4.24}$$

Hence the charge density ρ' has the properties of the Dirac function described in equations (4.19)–(4.21). In particular

$$\int_{v} \phi\rho'\, dv = \phi \tag{4.25}$$

so that the unit charge has a mutual energy with the system of charges ρ which is equal to the potential at the point at which the unit charge is located. In other words the unit charge provides a means of finding the potential of a system of charges.

The potential of the unit charge itself, given by equation (4.22), is called the free-space Green's function and is generally given the symbol G.

It follows from Green's theorem for an unbounded region that

$$\int_v \phi \nabla^2 \psi \, dv = \int_v \psi \nabla^2 \phi \, dv \qquad (4.26)$$

so that

$$\phi = \int_v G\rho \, dv \qquad (4.27)$$

which could have been obtained from equation (4.25) by transposing ϕ and G, and ρ' and ρ. In equation (4.25) we found the energy required to insert the unit charge into the field of charges ρ. In equation (4.27) we find the energy of inserting ρ into the field of the unit charge.

Equation (4.27) provides a useful tool for calculation. It is an application of the principle of superposition in which each element of charge density ρ contributes to the potential ϕ. The Green's function provides a means of taking account of the position of the element with respect to the field point at which ϕ is to be calculated.

For a bounded system we have from Green's theorem

$$\phi = \int_v G\rho \, dv + \int_s \left(G\sigma - \varepsilon_0 \frac{\partial G}{\partial n} \phi \right) ds. \qquad (4.28)$$

This involves a knowledge of the potential ϕ and its gradient $\partial\phi/\partial n = \sigma/\varepsilon_0$ at the surface s, and it is unlikely that this knowledge is available at the outset of a problem. Some simplification is possible by adding to the unit charge which is probing the system some surface layers or double layers of charge. We then have a modified Green's function

$$G = \frac{1}{4\pi\varepsilon_0 r} + \chi \qquad (4.29)$$

where χ is the potential of the charge layers and

$$\nabla^2 \chi = 0 \qquad (4.30)$$

within the volume v. It is possible by this means to put either G or $\partial G/\partial n$ equal to zero on s. An example of this method is given by the problem of a charge distribution ρ surrounded by an earthed conductor. We then have $\phi = 0$ on s, and if we choose $G = 0$ on s the surface term disappears and we have

$$\phi = \int_v G\rho \, dv. \qquad (4.31)$$

However, G is now no longer the free-space potential of a unit charge. It contains χ which in this case is the potential of the charge induced by the unit charge on an earthed conductor. The difficulty of finding χ is not much less than that of finding ϕ and the method is not particularly helpful. We have merely replaced the charge density ρ by the unit charge and have not got rid of the difficult boundary conditions.

More serious still is the objection that the method of Green's function focuses attention on the interaction of the system with a concentrated unit charge. Such a point probe collects the maximum information about the system which can be obtained because it finds the potential at every point. This information has to be paid for and mathematically the use of a Dirac function results in poor convergence of numerical processes. Often it is unnecessary to find the potential distribution because it is only the internal energy which is required. If that is so, a variational method is far more economical than the use of a Green's function. In numerical methods the boundary-element technique, the finite-difference scheme of calculation, and the highly discretized finite-element scheme correspond to the method of point Green's functions because they seek to obtain point information. Other 'moment' methods correspond to testing functions which are less concentrated and which as a result obtain average values for the mutual energy over a region. Variational methods use the energy of the system itself rather than testing the system by an auxiliary system. As would be expected there is much overlap between the methods because a part of a system can be used to probe another part. Underlying all the methods is Green's theorem which is an example of the principle of reciprocity or of mutual energy.

Before we leave this discussion of Green's theorem we note again that the energy it describes, whether it be the self-energy of a system or the mutual energy between two or more systems, is the equilibrium energy. For a small displacement of the system the change in energy will therefore be zero, so that we can use Green's theorem as a variational principle. The discovery of this aspect of Green's theorem is due to William Thomson (Lord Kelvin) and is known as Thomson's theorem. We shall discuss it under that name later in this chapter.

4.3. Magnetostatic systems and the vector form of Green's theorem

The sources of magnetostatic systems are of two kinds: magnetic dipoles and electric currents. If the sources are magnetic dipoles these can be replaced by an equivalent pole density and the energy can be written in terms of magnetic scalar potentials. This means that the entire mathematical description which we derived in the previous section in terms of electrostatics can be applied equally well to magnetostatic systems. The fact

that both electrostatics and magnetostatics have for their foundation stone
the inverse square law of force is most remarkable and leads to a beautiful
symmetry and duality in the description of electric and magnetic
phenomena.

In spite of these advantages there are important physical differences
because there are no magnetic conductors in nature and there is no free
magnetic polarity nor magnetic conduction current. Such quantities may
still be useful postulates in the mathematical formulation of problems but
they lack a physical basis. It may therefore be necessary to treat
magnetostatic problems in terms of electric current sources, although this
choice of sources introduces mathematical complication. If the reader feels a
distaste for the idea of magnetic poles, he may of course feel more confident
in the use of current sources but he should be aware of the extra work which
will result. It should also be remembered that Ampère's device of replacing a
current loop by a magnetic dipole layer or magnetic shell demonstrates the
equivalence of the two types of magnetic sources. This equivalence can be
applied only to systems in which both the 'go' and 'return' currents are
included, but this presents no difficulties to the discussion of the energy of
systems because as in electrostatic systems we can define the energy only if
the system is disconnected from the outside world and this means that there
must not be any net current flow. The possibility of using magnetic poles
and scalar potentials should therefore always be kept in mind in the
calculation of the energy of magnetostatic systems.

Nevertheless there are examples when it is desirable to start with current
sources and we shall now discuss how this can be done. It will be
remembered that the discussion of the previous section started with a
discussion of the divergence of the vector $\phi \nabla \psi$. Let us now consider the
divergence of the vector $\mathbf{P} \times (\nabla \times \mathbf{Q})$:

$$\int_v \nabla \cdot (\mathbf{P} \times \nabla \times \mathbf{Q})\, dv = \oint_s (\mathbf{P} \times \nabla \times \mathbf{Q}) \cdot d\mathbf{s} \tag{4.32}$$

whence

$$\int_v \{(\nabla \times \mathbf{P}) \cdot (\nabla \times \mathbf{Q}) - \mathbf{P} \cdot (\nabla \times \nabla \times \mathbf{Q})\}\, dv$$

$$= \oint_s (\mathbf{P} \times \nabla \times \mathbf{Q}) \cdot d\mathbf{s}. \tag{4.33}$$

Now let $\mathbf{P}$ represent the magnetic vector potential $\mathbf{A}$ defined by

$$\mathbf{V} \times \mathbf{A} = \mathbf{B} \tag{4.34}$$

$$\mathbf{V} \cdot \mathbf{A} = 0 \tag{4.35}$$

and let $\mathbf{Q}$ represent $(1/\mu)\mathbf{A}$, so that $\mathbf{V} \times \mathbf{Q}$ represents the magnetic field strength $\mathbf{H}$ defined by

$$\mathbf{V} \times \mathbf{H} = \mathbf{J} \tag{4.36}$$

$$\mathbf{B} = \mu\mathbf{H}. \tag{4.37}$$

Then equation (4.33) becomes

$$\int_v (\mathbf{B} \cdot \mathbf{H} - \mathbf{A} \cdot \mathbf{J})\, dv = \oint_s (\mathbf{A} \times \mathbf{H}) \cdot ds. \tag{4.38}$$

The surface magnetic field strength $\mathbf{H}$ can be replaced by a surface current of line density $\mathbf{I}$ which terminates the magnetic field beyond it. If we insert this in equation (4.38) we derive

$$\int_v (\mathbf{B} \cdot \mathbf{H} - \mathbf{A} \cdot \mathbf{J})\, dv = \oint_s (\mathbf{A} \cdot \mathbf{I})\, ds \tag{4.39}$$

and finally we can write

$$\tfrac{1}{2}\int \mathbf{A} \cdot \mathbf{J}\, dv + \tfrac{1}{2}\oint \mathbf{A} \cdot \mathbf{I}\, ds = \tfrac{1}{2}\int \mathbf{B} \cdot \mathbf{H}\, dv. \tag{4.40}$$

This should be compared with equation (4.6). As in that equation we have the assembly work on the left-hand side and the field energy on the right-hand side. The current density $\mathbf{I}$ is equal to the surface magnetic field strength $\mathbf{H}$ and therefore provides the return current for the currents described by $\mathbf{J}$. In this definition of the system we use $\mathbf{J}$ and $\mathbf{I}$ as the sources of the field. Alternatively we can regard the vector potential as the source of the field and write

$$-\tfrac{1}{2}\int \frac{1}{\mu}\mathbf{A} \cdot (\mathbf{V} \times \mathbf{V} \times \mathbf{A})\, dv + \tfrac{1}{2}\oint \frac{1}{\mu}(\mathbf{A} \times \mathbf{V} \times \mathbf{A}) \cdot ds = \tfrac{1}{2}\int \mathbf{B} \cdot \mathbf{H}\, dv \tag{4.41}$$

which is the analogue of equation (4.7)

The uniqueness of the energy of a current system can be examined as follows. Suppose there are two possible values of $\mathbf{A}$ such that

$$\mathbf{V} \times \mathbf{V} \times \mathbf{A}_1 = \mathbf{V} \times \mathbf{V} \times \mathbf{A}_2 = -\mu\mathbf{J}. \tag{4.42}$$

Apply equation (4.33) to the difference $\mathbf{C} = \mathbf{A}_1 - \mathbf{A}_2$. Then $\mathbf{V} \times \mathbf{V} \times \mathbf{C} = 0$ and

$$\int_v \frac{1}{\mu} (\mathbf{V} \times \mathbf{C})^2 \, dv = \int_s \frac{1}{\mu} (\mathbf{C} \times \mathbf{V} \times \mathbf{C}) \cdot d\mathbf{s}. \tag{4.43}$$

The surface term is zero if either $\mathbf{A}$ or $\mathbf{H}$ are specified on the surface. This, however, is more than is necessary as only the tangential components of $\mathbf{A}$ or $\mathbf{H}$ need be specified. This can be shown as follows:

$$\int_s (\mathbf{C} \times \mathbf{V} \times \mathbf{C}) \cdot d\mathbf{s} = \int_s (\mathbf{C} \times \mathbf{V} \times \mathbf{C}) \cdot \hat{\mathbf{n}} \, ds$$

$$= \int_s \mathbf{C} \cdot \{(\mathbf{V} \times \mathbf{C}) \times \hat{\mathbf{n}}\} \, ds$$

$$= \int_s (\mathbf{V} \times \mathbf{C}) \cdot (\hat{\mathbf{n}} \times \mathbf{C}) \, ds. \tag{4.44}$$

Hence if either C_t or $(\mathbf{V} \times \mathbf{C})_t$ are zero on s the surface term is zero. This is also true if C_t is constant on s, because then

$$C_t \int_s (\mathbf{V} \times \mathbf{C})_t \, ds = C_t \int_s (\mathbf{V} \times \mathbf{C}) \times \hat{\mathbf{n}} \, ds$$

$$= C_t \int \mathbf{V} \times \mathbf{V} \times \mathbf{C} \, dv = 0.$$

This implies by reference to equation (4.43) that within v

$$\mathbf{V} \times \mathbf{C} = 0 \tag{4.45}$$

and

$$\mathbf{A}_1 - \mathbf{A}_2 = \text{constant} \tag{4.46}$$

so that the magnetic field is uniquely specified and the field energy of the system is uniquely specified. The assembly work can be made equal to the field energy by choosing the arbitrary constant in equation (4.46) to be zero.

The mutual energy of two systems can be considered by putting $\mathbf{P} = \mathbf{A}$ and $\mathbf{Q} = \mathbf{A}'$ in equation (4.33) and then putting $\mathbf{P} = \mathbf{A}'$, $\mathbf{Q} = \mathbf{A}$ and subtracting. This leads to the vector analogue of Green's theorem as

$$\int_v \{\mathbf{A} \cdot (\mathbf{\nabla} \times \mathbf{\nabla} \times \mathbf{A}') - \mathbf{A}' \cdot (\mathbf{\nabla} \times \mathbf{\nabla} \times \mathbf{A})\} \, dv$$

$$= \oint_s (\mathbf{A}' \times \mathbf{\nabla} \times \mathbf{A} - \mathbf{A} \times \mathbf{\nabla} \times \mathbf{A}') \cdot d\mathbf{s} \qquad (4.47)$$

which should be compared with equation (4.15), noting the difference in the signs of the surface terms. If we substitute the currents we obtain

$$\int_v \mathbf{A} \cdot \mathbf{J}' \, dv + \oint_s \mathbf{A} \cdot \mathbf{I}' \, ds = \int_v \mathbf{A}' \cdot \mathbf{J} \, dv + \oint_s \mathbf{A}' \cdot \mathbf{I} \, ds. \qquad (4.48)$$

This should be compared with equation (4.17). It states that the mutual magnetostatic energy of two current systems $\mathbf{J}$, $\mathbf{I}$ and $\mathbf{J}'$, $\mathbf{I}'$ can be derived either by inserting the primed sources into the field of the unprimed ones or *vice versa*. Once again we are considering the equilibrium energy of a system which may contain many elements, so that this energy furnishes us with a variational principle.

There are many applications of the vector Green's theorem. A typical one is the calculation of mutual inductance where it is possible that one side of equation (4.48) provides an easier calculation than the other side. The primed quantities can also be used as a probe of the system of unprimed quantities. A particular probe is a small current of unit strength which has a Green's function for its vector potential. However, many other probes are possible and the choice is in the hands of the analyst.

4.4. Earnshaw's theorem

In our discussion of the energy of electrostatic and magnetostatic systems we have frequently mentioned the assembly work and have shown that this work is equal to the field energy. The notion of assembly work draws attention to some process by which the system is assembled and various processes are possible. Thus electric charges can be separated by chemical batteries or by diffusion processes which occur when different materials, for example p- and n-type semi-conductors, are in contact. Other possible processes are the thermoelectric and the photoelectric effects. Perhaps the most important processes are those in which mechanical work is done, as in the generation of electric current in electromagnetic machines of which there are many kinds or the generation of electric charge by electrostatic generators.

These various processes may not enter explicitly into our discussion, but they form the background for it. Without them there would be no energy and without the forces associated with them there would be no possibility of equilibrium. The importance of the associated mechanical forces will become particularly marked when we discuss the variation of the energy about the equilibrium position later in this chapter.

One particular aspect of the general statement that electric forces form a subsystem of the total forces and cannot be treated in isolation is discussed in this section. Earnshaw's theorem states that a charged body placed in an electric field cannot rest in stable equilibrium under the influence of the electric forces alone.

Let the charged body be described by A and let us suppose that at first all the charges on A and all other charges contributing to the electric field are fixed in their various bodies. Consider a particle of charge e located on A and let ϕ be the potential at that point. Then the potential energy of the charge e is $w = \phi e$. Also ϕ obeys Laplace's equation which can be written in rectangular co-ordinates

$$\frac{\partial^2 \phi}{\partial x^2} + \frac{\partial^2 \phi}{\partial y^2} + \frac{\partial^2 \phi}{\partial z^2} = 0. \tag{4.49}$$

Hence

$$\frac{\partial^2 w}{\partial x^2} + \frac{\partial^2 w}{\partial y^2} + \frac{\partial^2 w}{\partial z^2} = 0. \tag{4.50}$$

Consider now the other charges of A which can be typified by $w' = \phi' e'$. The potential ϕ' can be split into two parts, one depending on the potential ϕ and the other on the potential difference between ϕ' and ϕ. If the body A moves the second part will not change, so that ϕ' will obey equation (4.49). Hence, if we put

$$W = \sum w' \tag{4.51}$$

we have

$$\frac{\partial^2 W}{\partial x^2} + \frac{\partial^2 W}{\partial y^2} + \frac{\partial^2 W}{\partial z^2} = 0. \tag{4.52}$$

Now for stable equilibrium it is necessary that the second variation of the energy is positive, which implies that

$$\frac{\partial^2 W}{\partial x^2} > 0, \qquad \frac{\partial^2 W}{\partial y^2} > 0, \qquad \frac{\partial^2 W}{\partial z^2} > 0. \tag{4.53}$$

However, this is impossible by equation (4.52). Hence A cannot be in stable equilibrium. Here we have only considered translatory motion. The possibility of rotation merely reinforces the argument.

Now allow the charges of the system to move on their various conductors. They will move into positions of stable equilibrium which make the conductors into equipotentials. Hence the energy of the system will be decreased by such motion. However, a decrease of potential energy further increases the instability of the system of bodies such as A. Hence the theorem is proved.

It will be noticed that Earnshaw's theorem deals with potential energy only. In systems possessing both kinetic and potential energy, such as the gravitational system of the sun and its planets, stable equilibrium is possible. In electromagnetism it is also possible to suspend objects in a stable position by the interaction of potential and kinetic energy. Various levitation systems are examples of such energy combinations. Such arrangements are not easy to design and the difficulties are illustrated by the search for suitable geometries to contain a plasma suitable for fusion reactors.

It is therefore important to notice that in almost all electromagnetic devices stability has to be assured by constraints which are not electromagnetic in the sense that they do not fit into the system of forces described by Maxwell's equations. In variational methods Earnshaw's theorem is a reminder that energy can be exchanged with an electromagnetic system by various kinds of other forces, amongst which mechanical forces are the most important and easiest to envisage. This means that the scope of variational methods is greatly increased, as we shall see when we discuss Thomson's theorem.

4.5. Thomson's theorem

The application of variational methods to electricity and magnetism was first suggested by William Thomson, who later became Lord Kelvin. His guiding thought was that of the universality of the concept of energy and he interpreted Green's theorem as a statement about energy. He first saw Green's essay in 1845 when he was 20 years old and three years later published a paper which embodies his famous theorem.† He stated this theorem in very general terms as follows:

'Given the equation

$$\mathbf{V} \cdot (\alpha^2 \mathbf{V} \phi) = 4\pi\rho \tag{4.54}$$

where α is a real function of the coordinates and ρ is zero outside a finite closed surface, there exists a unique function ϕ of the coordinates.'

Although no examples are quoted Thomson associates this distribution of ϕ with the equilibrium state of the system and concludes his short paper by stating that his theorem has very important applications in the theories of

---

† *Cambridge and Dublin Mathematical Journal*, February 1848.

heat, electricity, magnetism and hydrodynamics. Much which can be inferred from his other writings is left unsaid in this paper. For instance in his diary he notes on 8 April 1845 that electrical energy is the product of charge and potential and that it has a minimum when the potential is constant. In the same passage he suggests that electric intensity should be measured in pounds per square inch rather than in the arbitrary divisions on the scale of an electrometer. The outstanding contribution made by the paper is the method of the proof in which he applies the calculus of variations to a functional which has a unique minimum value. Thomson's chief concern is not with the solution of the differential equation as such, but with an energy functional which has a minimum value if that equation is satisfied. These ideas, especially in relation to kinetic energy, were greatly expanded in Thomson and Tait's *Treatise on natural philosophy*, paragraphs 311, 312 and 316–319.

The full development of Thomson's theorem in its application to electricity is due to Maxwell (*Electricity and Magnetism*, Vol. 1, Part 1, Chapter 4). Since Thomson's theorem is central to the methods described in this book, we shall examine Maxwell's account in detail. He first considers the energy of a system enclosed by a surface on which there is an assigned distribution of potential $\bar{\phi}$. There are no charges within the volume bounded by the surface. By Green's theorem the energy can be written as

$$W_\phi = \tfrac{1}{2} \int_v \varepsilon E^2 \, dv = \tfrac{1}{2} \int_v \varepsilon [\nabla \phi]^2 \, dv. \tag{4.55}$$

Consider now the possibility of varying the potential distribution from ϕ to ϕ' within the volume, but let both ϕ and ϕ' take the assigned value $\bar{\phi}$ on the surface. Also let ϕ satisfy Laplace's equation within the volume. Consider the change in the energy as ϕ is replaced by ϕ':

$$W_{\phi'} - W_\phi = \tfrac{1}{2} \int \varepsilon ([\nabla \phi']^2 - [\nabla \phi]^2) \, dv$$

$$= \tfrac{1}{2} \int \varepsilon [\nabla \phi' - \nabla \phi]^2 \, dv + \int \varepsilon \nabla \phi \cdot (\nabla \phi' - \nabla \phi) \, dv. \tag{4.56}$$

The second term on the right-hand side can be transformed by the divergence theorem to give

$$\oint \varepsilon (\phi' - \phi) \frac{\partial \phi}{\partial n} \, ds - \int \varepsilon (\phi' - \phi) \nabla^2 \phi \, dv.$$

However, $\phi' - \phi = 0$ on s and $\nabla^2 \phi = 0$ in v so that this term goes to zero and equation (4.56) becomes

$$W_{\phi'} - W_{\phi} = \tfrac{1}{2} \int \varepsilon [\nabla\phi' - \nabla\phi]^2 \, dv. \tag{4.57}$$

Hence $W_{\phi'} > W_{\phi}$ unless $\nabla\phi' = \nabla\phi$ in v, i.e. unless $\phi' - \phi = $ constant in v. Moreover, since $\phi' - \phi = 0$ on s, the constant must itself be zero. Hence W_{ϕ} is the unique minimum value of the electrostatic energy of the system.

A case of particular interest is the system surrounded by an exterior surface s, on which $\bar{\phi} = 0$, and a set of interior surfaces $s_1, s_2, \ldots$, on which $\bar{\phi} = \bar{\phi}_1, \bar{\phi}_2, \ldots$. The energy of such a system takes on a unique minimum value.

It should be noticed that this theorem describes a variational method in which the potential ϕ is kept constant at its assigned distribution on the surface but is subjected to an arbitrary variation in the volume. The variation of the potential also affects its gradient, i.e. the electric field. However, the variation is not applied to the charge density on the surfaces nor to the density of charge, if any, in the volume. It should also be noticed that the behaviour of the energy functional W_{ϕ} as a minimum value implies that the first variation of W_{ϕ} at equilibrium is zero and that the second variation is positive. The energy varies as the square of the variation of the field and is thus very insensitive to small variations.

Consider next a system defined by charges on a set of conductors. As in the previous example we assume an exterior surface s, on which $\bar{\phi} = 0$, and set of interior surfaces $s_1, s_2, \ldots$, on which there are charges $\bar{Q}_1, \bar{Q}_2, \ldots$. The potentials of these surfaces are constant but are not otherwise specified. Let the field be expressed in terms of the electric flux vector $\mathbf{D}$. Then for each interior surface we have

$$- \oint_{s_1, s_2, \ldots} D_n \, ds = \bar{Q}_1, \bar{Q}_2, \ldots \tag{4.58}$$

where the negative sign arises from the fact that we are dealing with interior surfaces. We also have $\nabla \cdot \mathbf{D} = 0$ if there is no volume distribution of charge.

Consider the electrostatic energy of the system in terms of $\mathbf{D}$:

$$W_D = \tfrac{1}{2} \int (D^2/\varepsilon) \, dv. \tag{4.59}$$

Suppose now that the flux density $\mathbf{D}$ is varied to $\mathbf{D}'$; then the change of energy is given by

$$W_{D'} - W_D = \tfrac{1}{2} \int \frac{1}{\varepsilon} (D'^2 - D^2) \, dv$$

$$= \tfrac{1}{2} \int \frac{1}{\varepsilon} |\mathbf{D}' - \mathbf{D}|^2 \, dv + \int \frac{1}{\varepsilon} \mathbf{D} \cdot (\mathbf{D}' - \mathbf{D}) \, dv. \qquad (4.60)$$

Since $\mathbf{D} = \varepsilon \mathbf{E} = -\varepsilon \nabla \phi$, we can transform the second term on the right-hand side by the divergence theorem to

$$-\oint \phi (D_n' - D_n) \, ds + \int \phi \nabla \cdot (\mathbf{D}' - \mathbf{D}) \, dv.$$

However, the potential on the conductors is constant so that

$$-\oint \phi (D_n' - D_n) \, ds = -\phi \oint (D_n' - D_n) \, ds = 0 \qquad (4.61)$$

since the charges on the conductors have assigned values. Also the absence of charge density in the volume ensures that $\nabla \cdot (\mathbf{D}' - \mathbf{D}) = 0$. Hence

$$W_{D'} - W_D = \tfrac{1}{2} \int \frac{1}{\varepsilon} [\mathbf{D}' - \mathbf{D}]^2 \, dv \qquad (4.62)$$

and $W_{D'} > W_D$, unless $\mathbf{D}' = \mathbf{D}$ in v when $W_{D'} = W_D$. Thus W_D is a unique minimum value of the energy which occurs when the conducting surfaces are equipotentials.

It will be noted that $W_D = W_\phi$, as was of course expected. The variation of the energy can be carried out by altering the charge distribution σ on the conductors. The potential is taken as the constant potential at equilibrium. Maxwell drew attention to the fact that W_ϕ corresponds to a capacitative system typified by the capacitance

$$C = 2W_\phi / \phi^2 \qquad (4.63)$$

whereas W_D corresponds to a capacitance

$$C = \tfrac{1}{2}(Q^2 / W_D). \qquad (4.64)$$

Since both W_ϕ and W_D are minima, the method gives a minimum for the capacitance when ϕ is varied and a maximum when $\mathbf{D}$ is varied. We shall revert to this useful property in the next chapter. If we look back to the last chapter (§ 3.3.1) we notice that Thomson's theorem is a special example of the Lagrangian method applied to electrostatic problems. The variation of W_ϕ in equation (4.57) corresponds to equation (3.51) and the variation of W_D in equation (4.62) corresponds to equation (3.49).

4.6. Physical processes underlying Thomson's theorem

The physical basis of Thomson's theorem lies in the behaviour of systems subject to small disturbances from their equilibrium state. This is of course analogous to Lagrange's mechanical investigations. The disturbances are written as mathematical variations of the parameters describing the system and these variations are subject to constraints which can be introduced by the method of Lagrange's multipliers which was discussed in § 2.2. In this chapter we are concerned with a physical rather than a mathematical description and we now discuss some of the possible physical processes which underly Thomson's theorem. It is extremely helpful if we can picture a physical process which accompanies a particular method of calculation because this gives a qualitative understanding of such methods. The physical understanding enables us to choose appropriate mathematical tools.

We have already touched on one physical variation at the end of the previous section when we mentioned a displacement of charge density on a conductor. Let us now enlarge this discussion.

Consider first the variation of the energy W_ϕ of a system of charged conductors which have assigned values of potential. Maxwell postulated a variation of the potential ϕ within the volume of the system. How can this be brought about?

Let us insert an uncharged conductor into the system and notice how this affects the energy. Let the volume of the inserted conductor be v_0 and let the volume of the system be $v = v_1 + v_0$. Let us write the change of energy in terms of the changes of potential and electric field:

$$W_{\phi'} - W_\phi = \tfrac{1}{2} \int_{v_1} \varepsilon E'^2 \, dv - \tfrac{1}{2} \int_v \varepsilon E^2 \, dv \qquad (4.65)$$

since there is zero field within the volume v_0 because it is a conductor. Hence

$$W_{\phi'} - W_\phi = \tfrac{1}{2} \int_{v_1} \varepsilon(E'^2 - E^2) \, dv - \tfrac{1}{2} \int_{v_0} \varepsilon E^2 \, dv$$

$$= \tfrac{1}{2} \int_{v_1} \varepsilon[\mathbf{E'} - \mathbf{E}]^2 \, dv + \int_{v_1} \varepsilon\mathbf{E} \cdot (\mathbf{E'} - \mathbf{E}) \, dv$$

$$- \tfrac{1}{2} \int_{v_0} \varepsilon E^2 \, dv. \qquad (4.66)$$

Consider the second term

$$\int_{v_1} \varepsilon \mathbf{E} \cdot (\mathbf{E}' - \mathbf{E})\, dv = - \int_{v_1} \mathbf{D} \cdot \mathbf{V}(\phi' - \phi)\, dv$$

$$= - \oint_s (\phi' - \phi) D_n\, ds$$

$$+ \int_{v_1} (\phi' - \phi)\mathbf{V} \cdot \mathbf{D}\, dv. \tag{4.67}$$

The surface integral has to be taken over the original surfaces and also over s_0 which surrounds v_0. The former have assigned potentials so that $\phi' - \phi = 0$. The latter contributes

$$- \oint_{s_0} (\phi' - \phi) D_n\, ds = - \phi' \oint_{s_0} D_n\, ds + \oint_{s_0} \phi D_n\, ds.$$

The first term is zero because the inserted conductor is uncharged and the second is

$$- \int_{v_0} \varepsilon E^2\, dv$$

since

$$\mathbf{V} \cdot \mathbf{D} = 0.$$

Substituting in equation (4.66) we obtain

$$W_{\phi'} - W_\phi = \tfrac{1}{2} \int_{v_1} \varepsilon [\mathbf{E}' - \mathbf{E}]^2\, dv - \tfrac{3}{2} \int_{v_0} \varepsilon E^2\, dv. \tag{4.68}$$

Let us now specify that the inserted conductor is to be in the form of a very thin lamina of negligible volume. We then have

$$W_{\phi'} - W_\phi = \tfrac{1}{2} \int_v \varepsilon [\mathbf{E}' - \mathbf{E}]^2\, dv \tag{4.69}$$

which is the same as equation (4.57) and states that W_ϕ is the unique minimum value of the system energy.

We note that mechanical work has to be done to insert the thin conductor into the system and it is this work which increases the system energy. There is no energy interchange with the charged conductors because we have not varied the charges or the potentials. If the reader has doubts about this because the potentials are disturbed by the insertion of the thin conductor, these doubts can be resolved by devising a means of adjusting the potentials to bring them back to their assigned values while keeping the charges constant.

It is interesting to notice why the inserted conductor must be thin. In the variational process we have specified that the potential is changed from ϕ to ϕ' but that the charge distribution is unchanged. When a conductor is inserted into the system charge will be induced on its surface. However, if the conductor is thin its opposite sides will be so close together that the induced charges will cancel because they are of equal and opposite sign. Such a conductor is a double layer of charge, and a double layer alters the potential without introducing a volume charge.

Let us now investigate the possibility of altering the flux density $\mathbf{D}$ while holding ϕ constant. Consider the insertion of a piece of dielectric material into the system. Let this piece have volume v_0 enclosed by a surface s_0 and let it have the property of excluding field from its volume. We shall discuss what we mean by this shortly:

$$
W_{D'} - W_D = \tfrac{1}{2} \int_{v_1} \frac{1}{\varepsilon} D'^2 \, dv - \tfrac{1}{2} \int_{v_1 + v_0} \frac{1}{\varepsilon} D^2
$$

$$
= \tfrac{1}{2} \int_{v_1} \frac{1}{\varepsilon} [\mathbf{D}' - \mathbf{D}]^2 \, dv + \int_{v_1} \frac{1}{\varepsilon} \mathbf{D} \cdot (\mathbf{D}' - \mathbf{D}) \, dv - \tfrac{1}{2} \int_{v_0} \frac{D^2}{\varepsilon} \, dv.
$$

$$(4.70)$$

However,

$$
\int_{v_1} \frac{1}{\varepsilon} \mathbf{D} \cdot (\mathbf{D}' - \mathbf{D}) \, dv = \int_{v_1} -\nabla \phi \cdot (\mathbf{D}' - \mathbf{D}) \, dv
$$

$$
= -\oint \phi (D_{n'} - D_n) \, ds
$$

$$
+ \int_{v_1} \phi \nabla \cdot (\mathbf{D}' - \mathbf{D}) \, dv.
$$

$$(4.71)$$

As before there is no volume charge density in v so that $\mathbf{V} \cdot (\mathbf{D}' - \mathbf{D}) = 0$. The surface term consists of the surfaces on which there are fixed charges so that on these

$$- \int \phi(D_n' - D_n)\, ds = -\phi \int (D_n' - D_n)\, ds = 0$$

and there is also the surface s_0. For this surface we have

$$- \oint_{s_0} \phi(D_n' - D_n)\, ds = - \oint_{s_0} \phi D_n'\, ds + \int_{v_0} \left(\phi \mathbf{V} \cdot \mathbf{D} - \frac{1}{\varepsilon} D^2 \right) dv$$

$$= - \oint_{s_0} \phi D_n'\, ds - \int_{v_0} \frac{1}{\varepsilon} D^2\, dv. \qquad (4.72)$$

Substituting equation (4.72) in (4.70) we obtain

$$W_{D'} - W_D = \tfrac{1}{2} \int_{v_1} \frac{1}{\varepsilon} [\mathbf{D}' - \mathbf{D}]^2\, dv$$

$$- \oint_{s_0} \phi D_n'\, ds - \tfrac{3}{2} \int_{v_0} \frac{1}{\varepsilon} D^2\, dv. \qquad (4.73)$$

We now specify that the inserted material excludes the field from its interior by reducing the normal component of field to zero, i.e. $D_{n'} = 0$ on s_0. Physically this implies a surface charge. We have in mind a flux barrier, and it may clarify this idea to think of the flux as a current flow in which case the barrier would be an insulating body.

Finally we specify that the inserted body shall be in the form of a thin sheet, so that v_0 is negligible. Then

$$W_{D'} - W_D = \frac{1}{2} \int_v \frac{1}{\varepsilon} [\mathbf{D}' - \mathbf{D}]^2\, dv \qquad (4.74)$$

which is the same as equation (4.62) and ensures that W_D is the unique minimum energy. Thus in this variational process we introduce thin sheets of a material which excludes flux from its interior. Work has to be done to insert such flux barriers into the system and this work increases the internal energy. Again there is no other energy interchange because the assigned charges remain constant and their potentials are not varied. It is important to notice that the inserted material must have negligible volume if the potential and its gradient are to be unchanged. A body of finite volume would impose a change on the tangential field as well as on the normal field at its surface. The thin sheets which we envisage consist effectively of single layers of charge. Such layers do not affect the tangential field. Moreover, since our layers are such that they do not admit a normal component of field, they do not affect the potential.

The two variational processes which we have described are of fundamental importance to our method of calculation. The field is varied either by inserting conducting sheets or sheets which are flux barriers. These two types of insertion are the only ones which make it possible to vary either the potentials or the flux independently of each other. The application of these methods extends to all static vector fields and not only to electrostatics. Mathematically the conducting sheets affect the curl of the electric field and the flux barriers affect the divergence of the field. Admissible variations of the field therefore have to be variations of the curl only or of the divergence only. Consideration of the derivation of equations (4.69) and (4.74) will show that the argument holds if there are volume charges of constant strength in the volume of the charges of the system. In considering W_ϕ' $- W_\phi$ we can write

$$\int_{v_1} \varepsilon \mathbf{E} \cdot (\mathbf{E}' - \mathbf{E}) \, dv = \int_{v_1} -\nabla\phi \cdot (\mathbf{D}' - \mathbf{D}) \, dv$$

$$= -\oint \phi(D_n' - D_n) \, ds + \int_{v_1} \phi \nabla \cdot (\mathbf{D}' - \mathbf{D}) \, dv. \tag{4.75}$$

The second term is zero because $\nabla \cdot \mathbf{D}' = \nabla \cdot \mathbf{D}$ in v_1. The first term includes the charged surfaces and also the surface s_0. On the former the charge distribution is constant so that $D_n' = D_n$ and over s_0 there is no total charge either in the varied or the undisturbed field. Hence all the terms are zero and equation (4.69) stands.

In considering $W_D' - W_D$ we also note that the argument depends on $\nabla \cdot (\mathbf{D}' - \mathbf{D}) = 0$ in v_1 which is still correct if $\nabla \cdot \mathbf{D}' = \nabla \cdot \mathbf{D} = \rho$ in v_1. We also have $\nabla \cdot \mathbf{D} = 0$ in v_0 if the charge layer is inserted into a region of free charge. If there is charge density in v_0 originally, then $\int_{v_0} \phi \nabla \cdot \mathbf{D} \, dv$ is still zero as $v_0 \to 0$. Hence equation (4.74) stands.

Another variation which is of interest occurs when we insert a charged body of charge density ρ and a volume which is finite. We then have

$$W_D' - W_D = \frac{1}{2} \int_v \frac{1}{\varepsilon} [\mathbf{D}' - \mathbf{D}]^2 \, dv + \int_v \frac{1}{\varepsilon} \mathbf{D} \cdot (\mathbf{D}' - \mathbf{D}) \, dv. \tag{4.76}$$

The second term is

$$\int_v \frac{1}{\varepsilon} \mathbf{D} \cdot (\mathbf{D}' - \mathbf{D})\, dv = -\int_v \nabla\phi \cdot (\mathbf{D}' - \mathbf{D})\, dv$$

$$= -\oint_s \phi(D_n' - D_n)\, ds$$

$$+ \int_v \phi \nabla \cdot (\mathbf{D}' - \mathbf{D})\, dv. \qquad (4.77)$$

The surface term is zero on all the charged surfaces, except for the surface which encloses the whole system, so we have

$$\int_s \phi(\sigma' - \sigma)\, ds + \int_v \phi(\rho' - \rho)\, dv$$

but this is the mutual energy of the system of potential ϕ and sources σ, ρ with a system of sources $\sigma' - \sigma$ and $\rho' - \rho$. The mutual energy is necessarily positive. Hence

$$W_D' - W_D = \frac{1}{2} \int_v \frac{1}{\varepsilon} [\mathbf{D}' - \mathbf{D}]^2\, dv + \int_v \phi(\rho' - \rho)\, dv$$

$$+ \oint_s \phi(\sigma' - \sigma)\, ds. \qquad (4.78)$$

Thus the energy is raised beyond the value given by the variation which changes $\mathbf{D}$ to $\mathbf{D}'$ by the insertion of thin sheets of charge and the increase of the energy depends on the first order of the variation. It should be noted that the insertion of volume charges alters the potential as well as the flux density because

$$\int_v \phi(\rho' - \rho)\, dv + \oint_s \phi(\sigma' - \sigma)\, ds = \int (\phi' - \phi)\rho\, dv + \oint (\phi' - \phi)\sigma\, ds.$$

$$(4.79)$$

Before we leave Thomson's theorem we consider alternative methods of applying the variation, in which there is an energy interchange involving the charged conductors.

Consider first a variational process in which the potentials are changed throughout the system, not only in the volume but also on the surfaces. Let the change density be held constant in the volume and on the surfaces. The difference between this specification and the previous one is that there is an assigned surface charge density instead of an assigned surface potential distribution. The variation in the volume is again carried out by the insertion of conducting sheets:

$$W_\phi - W_\phi' = \tfrac{1}{2} \int_v \varepsilon [\mathbf{E}' - \mathbf{E}]^2 \, dv$$

$$+ \int_v \varepsilon \mathbf{E}' \cdot (\mathbf{E} - \mathbf{E}') \, dv. \tag{4.80}$$

The second term can be written as

$$- \int \nabla \phi' \cdot (\mathbf{D} - \mathbf{D}') \, dv = - \oint \phi'(D_n - D_n') ds$$

$$+ \int \phi' \nabla \cdot (\mathbf{D} - \mathbf{D}') \, dv. \tag{4.81}$$

Both these terms are zero since the charge density is held constant. Hence for this variation $W_\phi > W_\phi'$ and W_ϕ is the unique maximum value of the energy. Hence the conducting sheets are drawn into the system which does work and thus lowers its internal energy. If the variation involves conductors of finite volume v_0 the energy is still further lowered by the amount $\tfrac{1}{2} \int_{v_0} \varepsilon E^2 \, dv$. Finally consider a variational process in which charges are introduced both in the volume and on the conductors at points which are assumed connected to sources of charge at constant potentials. We have

$$W_D - W_D' = \frac{1}{2} \int_v \frac{1}{\varepsilon} [\mathbf{D} - \mathbf{D}']^2 \, dv + \int_v \frac{1}{\varepsilon} \mathbf{D}' \cdot (\mathbf{D} - \mathbf{D}') \, dv. \tag{4.82}$$

The second term can be written as

$$\int_v \mathbf{D}' \cdot (\nabla \phi' - \nabla \phi) \, dv = \oint_s (\phi' - \phi) D_n' \, ds$$

$$+ \int_v (\phi' - \phi) \nabla \cdot \mathbf{D}' \, dv. \tag{4.83}$$

Since the charges are introduced at constant potentials, both these terms are zero and $W_D > W_D'$ so that W_D is the unique maximum value of the energy.

In summary, we have discussed seven possible processes of varying the energy of a system of charges.

(1) An increase of energy proportional to the second order of the variation of the field caused by the insertion of thin conducting sheets into parts of the system initially devoid of charge or conducting matter. During this process no energy is exchanged with the original charges which are held constant at constant potential. The variation is equivalent to altering the potential of the electric field distribution, keeping the charge density constant.

(2) An increase of energy proportional to the second order of the variation of the field caused by disturbing the charge density on the conductors.

(3) An increase of energy proportional to the second order of the variation of the field caused by inserting thin flux barriers into the system. During this process no energy is exchanged with the original charges which are held constant at constant potential. This variation is equivalent to altering the flux density of the electric field while holding the potential constant.

(4) A decrease of energy proportional to the second order of the variation of the field caused by inserting thin conducting sheets into the system and allowing the potentials of the original conductors to take on new values. In this process work is done by the system in drawing the sheets into itself. There is an accompanying decrease of the potential energy of the original charges.

(5) A decrease of energy proportional to the second order of the variation of the field caused by inserting thin flux barriers into the system while maintaining the original charges at their potentials and inserting volume charge only at constant potential. In this process work is done by the system in drawing the thin sheets of charge into itself. The energy is supplied by the sources which maintain the fixed potentials.

(6) An increase of energy proportional to the first order of the variation of the field caused by inserting charge having a finite volume. No energy is interchanged with the original charges. This process is equivalent to changing both the charge and the potential of the field distribution.

(7) A decrease of energy containing a second-order variational term and a first-order term embodying the loss of field energy caused by inserting conducting material not only in thin sheets but in a finite volume, energy being derived from the potential energy of the system. This process is equivalent to changing both the potential and the charge of the field distribution.

For numerical work the first five processes are better than the last two because they provide a closer estimate of the actual energy. Moreover, it is these processes which provide variational principles.

All the results of this section hold for magnetostatic as well as electrostatic fields. We shall use the results in both contexts in the next chapter.

4.7. Poynting's theorem

In § 4.3 we examined the energy of magnetostatic systems by means of the vector Green's theorem. The same theorem is useful in an examination of time-varying electromagnetic fields. We replace the magnetic vector potential by the electric field strength and examine the divergence of the vector $\mathbf{S} = \mathbf{E} \times \mathbf{H}$:

$$\mathbf{V} \cdot (\mathbf{E} \times \mathbf{H}) = \mathbf{H} \cdot \mathbf{V} \times \mathbf{E} - \mathbf{E} \cdot \mathbf{V} \times \mathbf{H}. \tag{4.84}$$

Maxwell's equations provide

$$\mathbf{V} \times \mathbf{E} = -\frac{\partial \mathbf{B}}{\partial t} \tag{4.85}$$

and

$$\mathbf{V} \times \mathbf{H} = \mathbf{J} + \frac{\partial \mathbf{D}}{\partial t} \tag{4.86}$$

so that

$$\int_v \mathbf{V} \cdot (\mathbf{E} \times \mathbf{H}) \, \mathrm{d}v = \oint_s (\mathbf{E} \times \mathbf{H}) \cdot \mathrm{d}s$$

$$= -\int_v \left(\mathbf{H} \cdot \frac{\partial \mathbf{B}}{\partial t} + \mathbf{E} \cdot \frac{\partial \mathbf{D}}{\partial t} + \mathbf{E} \cdot \mathbf{J} \right) \mathrm{d}v \tag{4.87}$$

whence

$$\oint_s \mathbf{S} \cdot \mathrm{d}s = -\int \left[\mathbf{E} \cdot \mathbf{J} + \frac{\mathrm{d}}{\mathrm{d}t} \left(\tfrac{1}{2}\mathbf{H} \cdot \mathbf{B} + \tfrac{1}{2}\mathbf{E} \cdot \mathbf{D} \right) \right] \mathrm{d}v \tag{4.88}$$

where $\mathbf{S} = \mathbf{E} \times \mathbf{H}$.

This statement is known as Poynting's theorem and was first given by J. H. Poynting in a paper published in 1884 under the title *On the transfer of energy in the electromagnetic field*. The vector $\mathbf{S}$ is known as Poynting's vector.

This theorem has a large number of applications and repays careful study.

We notice that the various terms have the dimensions of power rather than energy, so that the theorem deals with energy flow. On the right-hand side of equation (4.88) there are two terms which express the rate of decrease of the electric and magnetic field energies within the volume. We have already noticed in our discussion of electrostatic systems in § 4.1 that the field energy is equal to the internal energy of the system only if the system is isolated by a suitable surface layer. In the electromagnetic case there are two types of energy, electric and magnetic, and two types of surface layers are required. It is clear from the surface integral in equation (4.58) that the energy flow across the surface depends on both the tangential electric field and the tangential magnetic field. However, the specification of either an electric or a magnetic current can reduce either H or E to zero. The energy flow is then zero and the system is isolated. Thus the system defined by the surface is isolated from the rest of space if there is an electric current of line density $\mathbf{I} = \mathbf{H} \times \hat{\mathbf{n}}$ or a magnetic current of line density $\mathbf{I}_m' = -\mathbf{E} \times \hat{\mathbf{n}}$ on the surface of the system (see Fig. 4.3).

Fig. 4.3

The power flowing outwards across the surface can be written in terms of these currents as

$$\oint \tfrac{1}{2}(\mathbf{H} \cdot \mathbf{I}_m + \mathbf{E} \cdot \mathbf{I})\,\mathrm{d}s = \oint \tfrac{1}{2}[-\mathbf{H} \cdot (\mathbf{E} \times \hat{\mathbf{n}}) + \mathbf{E} \cdot (\mathbf{H} \times \hat{\mathbf{n}})]\,\mathrm{d}s$$

$$= \oint \tfrac{1}{2}[-\mathbf{H} \times \mathbf{E}) \cdot \hat{\mathbf{n}} + (\mathbf{E} \times \mathbf{H}) \cdot \hat{\mathbf{n}}]\,\mathrm{d}s$$

$$= \oint \mathbf{S} \cdot \hat{\mathbf{n}}\,\mathrm{d}s. \tag{4.89}$$

The factor $\tfrac{1}{2}$ arises from the fact that $\mathbf{H}$ and $\mathbf{E}$ are the fields just inside the surface. Since the fields are zero outside the surface, the fields at the surface are $\tfrac{1}{2}\mathbf{H}$ and $\tfrac{1}{2}\mathbf{E}$. It is interesting to notice that the tangential components of the external magnetic field are terminated by the electric surface current and that the normal component of the magnetic field is terminated by the magnetic surface current since normal magnetic field is related to the tangential electric field by Maxwell's equation (4.85) and therefore the termination of the tangential electric field also terminates the normal magnetic field. For an open surface the reduction of all three components of the magnetic field requires both the electric and the magnetic surface currents, but for a closed surface one type of current is sufficient. It should also be noted that the surface currents of zero thickness are mathematical rather than physical concepts. The existence of metallic conductors makes it possible to have finite electric current in a very thin sheet, but the absence of magnetic conductors means that appreciable thickness is required for a finite magnetic flux. This thickness is of the order of the skin depth which is associated with permeability, conductivity, and frequency. The termination of electromagnetic fields at a surface is therefore not physically possible, but where the skin depth is very small compared with the dimensions of the system the ideal and practical phenomena are not very different. In our discussion we are chiefly concerned with the mathematical definition of system energies and not with measurement. Hence the surface currents are to be thought of as tools for the purpose of calculation. Nevertheless there is no point in calculating physical parameters which cannot be measured because they have no independent existence. The present discussion is therefore useful in defining the limits of applicability of our method of calculation. An interesting case arises if we consider electromagnetic field distributions E_1, H_1 and E_2, H_2. By the use of equation (4.88) we have

$$\oint (E_1 \times H_2 - E_2 \times H_1) \cdot \hat{n} \, ds$$

$$= \int \left(-H_2 \cdot \frac{\partial B_1}{\partial t} - \sigma E_1 \cdot E_2 - E_1 \cdot \frac{\partial D_2}{\partial t} \right.$$

$$\left. + H_1 \cdot \frac{\partial B_2}{\partial t} + \sigma E_2 \cdot E_1 + E_2 \cdot \frac{\partial D_1}{\partial t} \right) dv \qquad (4.90)$$

For simple harmonic time variation we have therefore

$$\oint (E_1 \times H_2 - E_2 \times H_1) \cdot \hat{n} \, ds = 0 \qquad (4.91)$$

and we can use Poynting's theorem as a statement that the mutual energy of two systems can be calculated by combining the electric field of one system with the magnetic field of the other system. Consideration of equation (4.90) will show that the statement holds also for systems in which currents or electric fields are impressed by sources within the systems.

Equation (4.91), like any other system equation, applies to a closed surface. In applying the statement to the mutual action of two systems it may be necessary to enclose both systems by surfaces s_1 and s_2 respectively and then to enclose the whole by another surface s_3. In principle therefore we shall have to take into account the contribution of the surface integral over s_3. However, if this surface is taken at a suitably large distance only the radiation fields will have a finite magnitude. For such fields

$$\mathbf{E}_1 = -\hat{\mathbf{n}} \times \mathbf{H}_1 \sqrt{\left(\frac{\mu}{\varepsilon}\right)} \qquad \mathbf{E}_2 = -\hat{\mathbf{n}} \times \mathbf{H}_2 \sqrt{\left(\frac{\mu}{\varepsilon}\right)}.$$

Hence over the distant surface s_3 we have

$$\oint_{s_3} (\mathbf{E}_1 \times \mathbf{H}_2 - \mathbf{E}_2 \times \mathbf{H}_1) \cdot \hat{\mathbf{n}} \, ds$$

$$= \oint_{s_3} \left| -\frac{\mu}{\varepsilon} (\hat{\mathbf{n}} \times \mathbf{H}_1) \cdot (\hat{\mathbf{n}} \times \mathbf{H}_2) + \frac{\mu}{\varepsilon} (\hat{\mathbf{n}} \times \mathbf{H}_2) \cdot (\hat{\mathbf{n}} \times \mathbf{H}_1) \right| ds$$

$$= 0. \tag{4.92}$$

Poynting's theorem is a statement about the conservation of energy and we should expect that it could be cast into the form of a variational principle. Equilibrium requires an adjoint system and we discussed this in §§ 3.3.4 and 3.3.5. It is, therefore, necessary to cast Poynting's theorem into adjoint form before it can be used as a variational principle. The easiest way of doing this is to assume a harmonic variation in time and to use complex notation as in equations (3.93) and (3.98), which are applicable to problems in which the displacement current is negligible, or as in equations (3.107) and (3.111) if the displacement current is included.

Closely related to the idea of equilibrium is the consideration of uniqueness, which we have already discussed in the context of Green's theorem in §§ 4.1 and 4.3.

We can write this theorem in adjoint form as

$$\oint (\mathbf{E} \times \mathbf{H}^*) \cdot \hat{\mathbf{n}} \, ds = \int \left| -\mathbf{H}^* \cdot \frac{\partial \mathbf{B}}{\partial t} - \mathbf{E} \cdot \left(\mathbf{J}^* + \frac{\partial \mathbf{D}^*}{\partial t} \right) \right| dv. \tag{4.93}$$

If $\mathbf{J}^* = \sigma \mathbf{E}^*$ is a current density induced in conducting material by the electric field $\mathbf{E}^*$, we can write $\mathbf{E} \cdot \mathbf{J}^* = \sigma E^2$. If the permeability μ and the permittivity ε are constant with time and the frequency is ω we have

$$\oint (\mathbf{E} \times \mathbf{H}^*) \cdot \hat{\mathbf{n}} \, ds = j\omega \int (\varepsilon \mathbf{E} \cdot \mathbf{E}^* - \mu \mathbf{H} \cdot \mathbf{H}^*) \, dv$$

$$- \int \sigma \mathbf{E} \cdot \mathbf{E}^* \, dv. \tag{4.94}$$

In the special case of linear polarization $\mathbf{E} \cdot \mathbf{E}^* = E^2$ and $\mathbf{H} \cdot \mathbf{H}^* = H^2$, but our argument does not require this condition.

Suppose now that there are two possible fields within the volume v and that these fields are denoted by the subscripts 1 and 2. Then we can write

$$\oint (\mathbf{E}_2 - \mathbf{E}_1) \times (\mathbf{H}_2 - \mathbf{H}_1)^* \cdot \hat{\mathbf{n}} \, ds$$

$$= j\omega \int \varepsilon (\mathbf{E}_2 - \mathbf{E}_1) \cdot (\mathbf{E}_2 - \mathbf{E}_1)^* \, dv$$

$$- j\omega \int \mu (\mathbf{H}_2 - \mathbf{H}_1) \cdot (\mathbf{H}_2 - \mathbf{H}_1)^* \, dv$$

$$- \int \sigma (\mathbf{E}_2 - \mathbf{E}_1) \cdot (\mathbf{E}_2 - \mathbf{E}_1)^* \, dv. \tag{4.95}$$

If either the tangential components of $\mathbf{E}_2$ and $\mathbf{E}_1$ or the tangential components of $\mathbf{H}_2$ and $\mathbf{H}_1$ are equal on the enclosing surface s, then the left side of equation (4.95) is zero. Taking the real part this means that $\mathbf{E}_2 = \mathbf{E}_1$ and for the imaginary part to be zero it means that $\mathbf{H}_2 = \mathbf{H}_1$. Thus the field in the volume is unique, if either tangential $\mathbf{E}$ or $\mathbf{H}$ is specified on the enclosing surface.

At first sight this is a surprising result because it suggests that the specification of either a magnetic or an electric current on the surface ensures uniqueness, whereas the energy flow depends on both types of source. Moreover the two types of source are not related to each other. The difficulty is resolved when we remember that the adjoint form of the theorem deals with average power flow. Hence if no power enters the volume through the surface, there can be no continuing average energy dissipation in conducting material within the volume and the internal fields decay to zero.

But if there is no dissipation within the volume uniqueness is not guaranteed unless both the tangential electric field and the tangential magnetic field are specified on the surface. For instance resonant cavity modes can be supported even if there is zero electric field on the surface of the cavity. The source of such modes is to be found in the wall currents and their associated charges. In addition modes are possible in which the surface integral of the average Poynting vector is zero, although neither the tangential electric field nor the tangential magnetic field is zero over a closed surface. Such modes are used in waveguides and transmission lines.

We conclude that uniqueness requires more information than equilibrium, because both the magnetic and the electric sources must be specified.

Equilibrium conditions are derived from a Lagrangian formulation, which by d'Alembert's principle gives different signs to the potential and kinetic energies. Conservation of energy and uniqueness are derived from a Hamiltonian formulation which gives the sum of the energies. The two ideas overlap completely only when there is a single type of energy.

Poynting's theorem can be used to lead to an extended form of Tellegen's theorem. We have noted that a system represented by network elements assumes the existence of a potential difference across the elements which completely defines the electric field. A second assumption in such a representation is that the current is uniquely defined at every instant. These two assumptions form the basis of Kirchhoff's voltage and current laws. In time-varying electromagnetic fields Kirchhoff's laws are not satisfied because the electric field has a non-conservative component and because the conduction current varies along the conductors.

However, the power and simplicity of the network concepts are such that they cannot be lightly dismissed. These are the guiding thoughts of the following discussion. Let us define the idea of a terminal pair, which has the property that the current entering by one terminal is always equal and opposite to the current leaving the other terminal. In circuit language this means that the terminal pair has negligible capacitance. Secondly let the terminal pair have the property that the potential difference between the terminals has a unique value which implies that the inductance of the terminal pair is negligible. Now impose the condition that no energy transfer takes place except at the terminal pairs.

We have

$$\oint_s (\mathbf{E} \times \mathbf{H}) \cdot \hat{\mathbf{n}}\, ds = \oint_s \left[\left(-\frac{\partial A}{\partial t} - \nabla\phi \right) \times \mathbf{H} \right] \cdot \hat{\mathbf{n}}\, ds. \qquad (4.96)$$

However,

$$\oint_s (-\nabla\phi \times \mathbf{H}) \cdot \hat{\mathbf{n}}\, ds = \int_v \nabla \cdot (-\nabla\phi \times \mathbf{H})\, dv$$

$$= \oint_s (\phi \nabla \times \mathbf{H}) \cdot \hat{\mathbf{n}}\, ds. \qquad (4.97)$$

Hence

$$\oint_s (\mathbf{E} \times \mathbf{H}) \cdot \hat{\mathbf{n}}\, ds = \oint_s \left(\phi \nabla \times \mathbf{H} - \frac{\partial A}{\partial t} \times \mathbf{H} \right) \cdot \hat{\mathbf{n}}\, ds$$

$$= \int_s \left(\phi \mathbf{J} + \phi \frac{\partial D}{\partial t} - \frac{\partial A}{\partial t} \times \mathbf{H} \right) \cdot \hat{\mathbf{n}}\, ds \qquad (4.98)$$

and if the capacitance and inductance terms are negligible

$$\oint_s (\mathbf{E} \times \mathbf{H}) \cdot \hat{\mathbf{n}} \, \mathrm{d}s = \oint_s \phi \mathbf{J} \cdot \hat{\mathbf{n}} \, \mathrm{d}s = \sum_i V_i I_i \qquad (4.99)$$

where V_i is the potential difference between the terminal pair carrying the current I_i. Tellegen's theorem $\sum V_i I_i = 0$ follows if there is no net energy flow through the surface s. This derivation of the theorem enlarges its scope because no assumptions need to be made about the internal structure of the system.

•   If the network has only two terminal pairs A and B we have

$$V_A I_A + V_B I_B = 0. \qquad (4.100)$$

In order to obtain a reciprocal relationship between the two terminal pairs we notice that the reciprocal form of Poynting's theorem is

$$\oint_s (\mathbf{E}_1 \times \mathbf{H}_2 - \mathbf{E}_2 \times \mathbf{H}_1) \cdot \hat{\mathbf{n}} \, \mathrm{d}s = 0 \qquad (4.101)$$

as long as the integral includes all the sources of the field. Hence

$$V_{A_1} I_{A_2} + V_{B_1} I_{B_2} = V_{A_2} I_{A_1} + V_{B_2} I_{B_1}. \qquad (4.102)$$

If

$$\begin{aligned} I_{A_2} = I_{B_1} = 0 \\ V_{B_1}/I_{A_1} = V_{A_2}/I_{B_2}. \end{aligned} \qquad (4.103)$$

Hence the mutual impedances of the two terminal pairs are equal.

In this discussion the surface s has been taken to enclose all the sources, but in some mutual impedance problems† it may be better to consider a surface surrounding only those sources connected to one of the terminal pairs. Consider first a set of surfaces which excludes all sources as shown in Fig. 4.4. The contribution of that part of the surface labelled s_0 can be made negligibly small. The complete surface can therefore be regarded as consisting of s_1 and s_2. Hence

$$\oint_{s_1 + s_2} (\mathbf{E}_1 \times \mathbf{H}_2 - \mathbf{E}_2 \times \mathbf{H}_1) \cdot \hat{\mathbf{n}} \, \mathrm{d}s = 0 \qquad (4.104)$$

---

† See A. L. Cullen and J. C. Parr, A new perturbation method for measuring microwave fields in free space, *Proc. Inst. Electr. Eng., Part B*, **6**, 836–844 (1955).

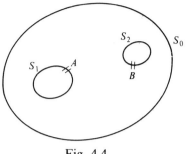

Fig. 4.4

and hence

$$\oint_{s_1}(\mathbf{E}_1 \times \mathbf{H}_2 - \mathbf{E}_2 \times \mathbf{H}_1) \cdot \hat{\mathbf{n}} \, ds = -\oint_{s_2}(\mathbf{E}_1 \times \mathbf{H}_2 - \mathbf{E}_2 \times \mathbf{H}_1) \cdot \hat{\mathbf{n}} \, ds. \quad (4.105)$$

Now let $\mathbf{E}_1$ and $\mathbf{H}_1$ be the fields associated with injecting a current of unit strength at A with B on open circuit. Similarly let $\mathbf{E}_2$ and $\mathbf{H}_2$ be the fields associated with injecting a current of unit strength at B with A on open circuit. The surface integral can be written

$$-\oint_{s_2}(\mathbf{E}_1 \times \mathbf{H}_2 - \mathbf{E}_2 \times \mathbf{H}_1) \cdot \hat{\mathbf{n}} \, ds = -(V_{B_1}I_{B_2} - V_{B_2}I_{B_1}). \quad (4.106)$$

But $I_{B_1} = 0$, $I_{B_2} = 1$ and $-V_{B_1} = Z_{AB}$ the mutual impedance. Hence

$$Z_{AB} = \oint_{s_1}(\mathbf{E}_1 \times \mathbf{H}_2 - \mathbf{E}_2 \times \mathbf{H}_1) \cdot \mathbf{n} \, ds. \quad (4.107)$$

The surface integral can be transformed into a volume integral and we then have

$$Z_{AB} = \int_{v} -\mathbf{E}_1 \cdot \mathbf{J}_2 \, dv. \quad (4.108)$$

Thus the electric field E_1 could be measured by a small dipole and would be proportional to the mutual impedance. This, like other applications of Poynting's theorem, can be cast into variational form.

This discussion of Poynting's theorem shows how the postulation of ideal terminal pairs enables a bridge to be built between circuit theory and field theory.

4.8. Equilibrium and uniqueness conditions in terms of the potentials

Poynting's theorem deals with the rate of change of energy rather than the energy itself. This means that we need an adjoint formulation before we can insert it into the Lagrangian scheme. The easiest way of achieving this is by the use of complex conjugates. We then deal with the average power flow and the average stored energy.

For non-dissipative systems this restriction can be removed by the use of the potentials. If $\mathbf{A}$ is the vector potential and ϕ is the scalar potential, the internal kinetic energy is $\frac{1}{2}\langle \mathbf{J}, \mathbf{A} \rangle$, and the internal potential energy $\frac{1}{2}\langle \rho, \phi \rangle$. In order to isolate the system we need a surface current $\mathbf{I}$ with its associated kinetic energy $\frac{1}{2}|\mathbf{I} \cdot \mathbf{A}|$ and a surface charge density with an associated potential energy $\frac{1}{2}|\sigma, \phi|$. These densities are of course connected by the continuity equation (1.32). The Lagrangian is given by the difference between the kinetic and potential energies. To obtain a variational principle this can be written in terms of coordinates and velocities which are to be varied, or alternatively it can be written in terms of the momenta and forces. The reader should now turn back to equations (3.125) and (3.129) to satisfy himself that these two equations are none other than the Lagrangian energy variations. Perhaps the method of § 3.3.5 is more straightforward than the method of assembling the various energy terms which is outlined in this section. But once the variational principle has been obtained it should be examined for its physical content.

The uniqueness conditions are also easily seen in terms of the potentials. Equation (1.39) and (1.42) show the potentials expressed in terms of their sources. If we postulate two different sets of potentials which have the same volume and surface sources, then the difference between these potentials have no sources at all for an isolated system. Thus the differences are identically equal to zero and the potentials are uniquely specified throughout the system. Hence the electric and magnetic fields are also unique.

4.9. Rayleigh's principle

In his classic work *The Theory of Sound* Lord Rayleigh used variational methods to obtain resonant frequencies and attenuation factors of vibrating systems. In the context of electromagnetism his work is of particular usefulness in its application to high-frequency systems, where the value of the frequency rather than the field energy itself is required. Rayleigh, Thomson, and Maxwell were contemporaries and kept in touch with each other, each acknowledging the influence of the other two, so that their work overlaps to a considerable extent. Rayleigh states that vibration in general may be considered as a periodic transformation of energy from the potential to the kinetic and from the kinetic to the potential forms. He notes that in a system without dissipation the average kinetic and potential energies are equal. This is equivalent to equating the Lagrangian of such a system to zero.

In our treatment of Thomson's theorem in § 4.6 we discussed variational processes which altered the value of the generalized co-ordinates and velocities and hence the energy in terms of the spatial co-ordinates. Now we note that these field quantities are functions of the frequency as well as of the spatial co-ordinates, if we are concerned with time-varying systems. We can still vary the system, but we now consider the variation in terms of the frequency.

As an example of the method consider the wave equation

$$\mathbf{V} \times \mathbf{V} \times \mathbf{E} - \omega^2 \mu\varepsilon\mathbf{E} = 0, \tag{4.109}$$

which can be obtained by combining equations (3.80), (3.81), (3.83), and (3.103) for the special case of $\mathbf{J} = 0$. A variational principle can be formed by writing

$$\langle (\mathbf{V} \times \mathbf{V} \times \mathbf{E} - \omega^2 \mu\varepsilon\mathbf{E}), \delta\mathbf{E}^* \rangle = 0. \tag{4.110}$$

This can be transformed into

$$\langle \mathbf{V} \times \mathbf{E}, \mathbf{V} \times \delta\mathbf{E}^* \rangle - |\delta\mathbf{E}^* \times (\mathbf{V} \times \mathbf{E}), \hat{\mathbf{n}}| - \langle \omega^2 \mu\varepsilon\mathbf{E}, \delta\mathbf{E}^* \rangle = 0, \tag{4.111}$$

and hence

$$\delta\{\langle |\mathbf{V} \times E|^2 \rangle - \omega^2 \langle \mu\varepsilon|E|^2 \rangle - |\mathbf{E}^* \times (\mathbf{V} \times \mathbf{E}), \hat{\mathbf{n}}|\} = 0. \tag{4.112}$$

If we put

$$\omega^2 = \frac{\int |\mathbf{V} \times \mathbf{E}|^2 \, dv - \oint |\mathbf{E}^* \times (\mathbf{V} \times \mathbf{E})| \cdot \hat{\mathbf{n}} \, ds}{\int \mu\varepsilon|E^2| \, dv} \tag{4.113}$$

this gives the stationary value of ω, because it can readily be shown that equation (4.113) is the necessary and sufficient condition for $\delta\omega = 0$. If the quantity to be varied in equation (4.112) is itself zero, then we have Rayleigh's condition of equality between the average potential and the average kinetic energy. It is shown by Berk[†] that μ and ε need not be scalar quantities, but can be tensor functions of the spatial co-ordinates as long as they are non-dissipative. If the trial function for E has zero tangential components on the enclosing surface, equation (4.113) can be simplified to give

$$\omega^2 = \frac{\int |\mathbf{V} \times \mathbf{E}|^2 \, dv}{\int \mu\varepsilon|E|^2 \, dv}, \tag{4.114}$$

† A. D. Berk, Variational Principles for Electromagnetic Resonators and Waveguides, *I.R.E. Transactions on Antennas and Propagation*, **AP-4**, 104–111, April 1956.

This expression is useful in determining the resonant frequencies of cavity resonators.

In some applications it is useful to find the resonant frequency in terms of both the electric and magnetic field. This can be done as follows.

We take the variational principles

$$\langle(\mathbf{V} \times \mathbf{H} - j\omega\varepsilon\mathbf{E}), \delta\mathbf{E}^*\rangle = 0 \qquad (4.115)$$

$$\langle(\mathbf{V} \times \mathbf{E} + j\omega\mu\mathbf{H}), \delta\mathbf{H}^*\rangle = 0 \qquad (4.116)$$

for a region devoid of sources. If we combine these we obtain

$$j\omega\langle\varepsilon\mathbf{E}, \delta\mathbf{E}^*\rangle + j\omega\langle\mu\mathbf{H}, \delta\mathbf{H}^*\rangle + \langle\delta\mathbf{H}^*, \mathbf{V} \times \mathbf{E}\rangle - \langle\delta\mathbf{E}^*, \mathbf{V} \times \mathbf{H}\rangle = 0 \qquad (4.117)$$

$$\delta\left\{\frac{j\omega}{2} \langle\varepsilon|E|^2 + \mu|H|^2\rangle + \tfrac{1}{2}\langle\mathbf{H}^*, \mathbf{V} \times \mathbf{E}\rangle - \tfrac{1}{2}\langle\mathbf{E}^*, \mathbf{V} \times \mathbf{H}\rangle\right\} = 0 \qquad (4.118)$$

which gives

$$\omega = \frac{j(\int \mathbf{H}^* \cdot \mathbf{V} \times \mathbf{E}\, dv - \int \mathbf{E}^* \cdot \mathbf{V} \times \mathbf{H}\, dv)}{\int \varepsilon|E|^2\, dv + \int \mu|H|^2\, dv}, \qquad (4.119)$$

However, this expression holds for a system only if no energy passes through the enclosing surface. This would be the case if the trial functions of **E** or **H** were given zero tangential values on the enclosing surface. More generally we must include the surface term to give

$$\omega = \frac{j(\int \mathbf{H}^* \cdot \mathbf{V} \times \mathbf{E}\, dv - \int \mathbf{E}^* \cdot \mathbf{V} \times \mathbf{H}\, dv - \oint (\mathbf{E} \times \mathbf{H}^*) \cdot \hat{\mathbf{n}}\, ds - \oint (\mathbf{E}^* \times \mathbf{H}) \cdot \hat{\mathbf{n}}\, ds)}{\int \varepsilon|E|^2\, dv + \int \mu|H|^2\, dv} \qquad (4.120)$$

It will be noted that this variational statement involves both **E** and **H**, i.e. both the co-ordinates and the momenta. This implies that we are using the Hamiltonian and not the Lagrangian energy and is confirmed by noting that the denominators of equations (4.119) and (4.120) give the sums of the potential and kinetic energies. The Hamiltonian expression does not allow a dual to be constructed by interchanging the co-ordinates and the momenta and in this respect the Lagrangian solution is preferable.

For example equation (4.114) can be interpreted as

$$\omega^2 = \frac{U}{X}. \qquad (4.121)$$

where $T = \omega^2 X$, T is the kinetic energy and U the potential energy. The stationary value of ω^2 is given if $T = U$.

The second variation of ω^2 is given by

$$\delta^2(\omega^2) = \delta^2\left(\frac{U}{X}\right). \tag{4.122}$$

An interchange of kinetic and potential energy will cause a change in the sign of this variation of ω^2 and will convert an upper bound to a lower bound of the solution, so that dual bounds can be obtained. The propagation constant of a transmission line can be obtained in a similar manner to the resonant frequency. Let this constant be denoted by γ and let the field vectors be typified by $\mathbf{E}(x, y)\, e^{-j\gamma z}$. From Maxwell's equations we have

$$\nabla \times \mathbf{E} + j\omega\mu\mathbf{H} = \gamma j\, \hat{\mathbf{a}}_z \times \mathbf{E} \tag{4.123}$$

$$\nabla \times \mathbf{H} - j\omega\varepsilon\mathbf{E} = \gamma j\, \hat{\mathbf{a}}_z \times \mathbf{H}. \tag{4.124}$$

Multiplying equation (4.123) by $\mathbf{H}^*$ and (4.124) by $\mathbf{E}^*$ and subtracting and integrating over the cross-section s we obtain a variational expression for γ as follows:

$$\gamma = \frac{\omega \int \varepsilon|E|^2\, ds + \omega \int \mu|H|^2\, ds + j \int \mathbf{E}^* \cdot \nabla \times \mathbf{H}\, ds - j \int \mathbf{H}^* \cdot \nabla \times \mathbf{E}\, ds.}{\int \mathbf{H}^* \cdot (\hat{\mathbf{a}}_z \times \mathbf{E})\, ds - \int \mathbf{E}^* \cdot (\hat{\mathbf{a}}_z \times \mathbf{H})\, ds}$$

$$\tag{4.125}$$

As before this expression requires either $\mathbf{E}$ or $\mathbf{H}$ to have zero tangential components at the circumference. If this is not the case the numerator must include the terms

$$j \oint (\mathbf{E} \times \mathbf{H}^*) \cdot d\mathbf{l} + j \oint (\mathbf{E}^* \times \mathbf{H}) \cdot d\mathbf{l}.\dagger \tag{4.126}$$

In this treatment we have assumed loss-free propagation. The scope of the method is extended to lossy transmission lines and waveguides in the excellent book by R. E. Collin, *Field Theory of Guided Waves*.

Further reading

A useful section on energy theorems is given by J. A. Stratton, *Electromagnetic Theory*, Chapter 2 (McGraw-Hill, 1941).

Maxwell's treatment of Thomson's theorem in his *Treatise on Electricity and Magnetism* is instructive.

---

† For an extended treatment see the paper by K. Marishita and N. Kumagai: 'Unified approach to the derivation of variational expression for electromagnetic fields,' *I.E.E.E. Trans,* **MTT 25**, No. 1, January 1977, pp. 34–40, as well as the paper by Berk. Our expressions (4.120) and (4.126) differ from both papers by including an additional boundary term, which in our view is required to make the equations consistent by including both the magnetic and electric surface sources.

Variational problems, particularly in regard to kinetic energy, are treated by Lord Kelvin and P. G. Tait *Treatise on Natural Philosophy* (Cambridge, 1896), but the treatment is highly abstract.

An excellent account of Poynting's theorem is given by G. D. Monteath, *Applications of the Electromagnetic Reciprocity Principle* (Pergamon, 1973).

Lord Kelvin's biography by S. P. Thompson (Macmillan, 1910) gives a detailed and fascinating account of Lord Kelvin and his contemporaries.

Lord Rayleigh, *The Theory of Sound*, Macmillan, London (Second Edition, 1894, reprinted, 1926).

A very full account of the application of variational methods to high-frequency electromagnetic phenomena is given by R. E. Collin, *Field Theory of Guided Waves*, McGraw-Hill (New York, 1960).

An interesting paper which shows how to enlarge the scope of the method is given by K. Marishita and N. Kumagai: 'Systematic Derivation of Variational Expressions for Electromagnetic and/or Acoustic Waves,' *I. E. E. E. Trans.*, **MTT 26**, 9, Sept. 1975, pp. 684–689.

5. Upper and Lower Bounds for Circuit Parameters

5.1. Potential and kinetic energy

In this section we remind ourselves of some results briefly mentioned in Chapter 2 (§ 2.1) where we discussed the principle of virtual work and showed that this principle is particularly well adapted for the examination of the equilibrium of systems. Newtonian mechanics deals with the equilibrium of the members of a system, be they mass points or rigid bodies. The principle of virtual work, on the other hand, deals with the equilibrium of the system as a whole by examining the variation of the energy of the system under an arbitrary small displacement or variation. Whereas the Newtonian equations examine the balance of the forces and torques, the method of virtual work proceeds by displacing the system and watching its behaviour. It examines the behaviour near equilibrium and can distinguish between stable and unstable equilibrium. For static systems in which the energy is conserved this energy depends only on the configuration and is called the potential energy. The condition of stable equilibrium is

$$\delta w \leqslant 0 \qquad (5.1)$$

which corresponds to

$$\delta U \geqslant 0 \qquad (5.2)$$

where U is the potential energy. The potential energy has a minimum at equilibrium. A typical example is a mass hanging from an inextensible string. When the mass is displaced the energy is increased and its value cannot lie below a convex surface as illustrated in Fig. 5.1. Equations (5.1) and (5.2) can be combined to show that the energy is conserved since

$$\delta w = -\delta U. \qquad (5.3)$$

We showed in Chapter 2 that this analysis could be extended to dynamic systems by using d'Alembert's principle of reversed effective forces. However, we also saw that it was not possible to derive such forces from a single energy expression. In other words the virtual work of the dynamic forces cannot be integrated. Of course we can and do associate an energy

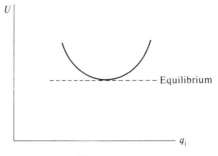

Fig. 5.1

with the motion, and this is the kinetic energy, but we cannot obtain the resultant force by differentiating the kinetic energy with regard to an arbitrary displacement. Thus the kinetic energy plays a different role from that of the potential energy.

The relationship of the two types of energy can be simplified by integrating the virtual work of the dynamic forces with respect to time. This integration is an integration by parts, which is a special case of the use of adjoint differential operators which were discussed in § 3.3.1. Two terms result from the integration: a term involving displacement along the direction of the motion and another term involving the variation of the kinetic energy with velocity. For a system of mass points we have

$$\int \delta w \, \mathrm{d}t = \int - \sum_i \frac{\mathrm{d}}{\mathrm{d}t} (m_i \mathbf{v}_i) \cdot \delta \mathbf{R}_i$$

$$= -\left[\sum_i m_i \mathbf{v}_i \cdot \delta \mathbf{R}_i \right] + \int \delta T \, \mathrm{d}t. \tag{5.4}$$

For the definite integral in the time interval from t_1 to t_2

$$\int_{t_1}^{t_2} \delta w \, \mathrm{d}t = -\left[\sum_i m_i \mathbf{v}_i \cdot \delta \mathbf{R}_i \right]_{t_1}^{t_2} + \int_{t_1}^{t_2} \delta T \, \mathrm{d}t. \tag{5.5}$$

If we arrange for the δR_i to be zero at $t = t_1$ and $t = t_2$, we have in the absence of potential energy

$$\int_{t_1}^{t_2} \delta w \, \mathrm{d}t = \int_{t_1}^{t_2} \delta T \, \mathrm{d}t. \tag{5.6}$$

However, the omission of the first term on the right-hand side of equation (5.5) applied to the definite integral only. During the motion δR_i is certainly not zero. Indeed in a purely kinetic system the energy balance demands such a term because both the incremental work done by the system and its incremental stored energy have the same sign. Thus there has to be a balancing energy which provides both the work and the kinetic energy. $\Sigma_i m_i \mathbf{v}_i$ can be regarded as a set of impulsive forces causing the motion $\mathbf{v}_i$. Changes in the kinetic energy must always be accompanied by such forces. In a potential energy system the potential energy can be changed into work without the intervention of such additional forces and there is therefore an important difference between the two types of energy. In particular we have for the equilibrium condition of equation (5.1) applied to equation (5.6) in a purely kinetic system that

$$\delta T \leqslant 0. \tag{5.7}$$

Hence T has a maximum value at equilibrium (see Fig. 5.2). Thus the maximum and minimum energy conditions which we have met in Chapters 3 and 4 can be explained by classifying the energies as kinetic or potential.

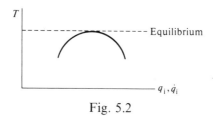

Fig. 5.2

5.2. Potential and kinetic energy formulation for the energy of electrostatic systems

In § 3.3.1 we discussed variational energy principles for electrostatic systems. We obtained four such principles and we shall now look at them again in the light of the discussion of the previous section. For the sake of completeness we shall include the surface terms which were omitted in § 3.3.1. We have

$$\delta\{[\sigma, \phi] + \langle \bar{\rho}, \phi \rangle - \tfrac{1}{2}\varepsilon \langle E^2 \rangle\} = 0 \tag{5.8}$$

where the variation is carried out in ϕ and $\mathbf{E}$,

$$\delta\left\{[\sigma, \bar{\phi}] + \langle \rho, \bar{\phi} \rangle - \frac{1}{2\varepsilon}\langle D^2 \rangle\right\} = 0 \tag{5.9}$$

where the variation is carried out in $\mathbf{D}$, σ and ρ,

$$\delta\left\{\frac{\varepsilon}{2}\langle E^2 \rangle\right\} = 0 \tag{5.10}$$

where $\mathbf{E}$ is varied, and finally

$$\delta\left\{\frac{1}{2\varepsilon}\langle D^2\rangle\right\} = 0 \qquad (5.11)$$

where $\mathbf{D}$ is varied. We notice that the first two expressions have the characteristics of kinetic energy and the second two of potential energy because in the former the second variation is negative and in the latter positive.

Since electrostatic systems would normally be thought of as possessing potential energy only it seems strange at first sight that it is possible to use a kinetic energy formulation. However, we are helped by our discussion of Thomson's theorem in Chapter 4 (§ 4.6). Consider first the variation of ϕ and $\mathbf{E}$. We discussed this in the context of inserting thin sheets of conducting material into the space between the charges. If the original charges and potentials were held constant, then work had to be done to insert the thin sheets into the system and its energy increased. Clearly this is a case of potential energy, as in equations (4.69) and (5.10). If, on the other hand, the potentials of the original charges were allowed to vary, the conducting sheets were drawn into the system which therefore did work on its environment. Energy was provided by the original charges, half of it increasing the field energy and half of it providing the work term. This corresponds to equations (4.80) and (5.8). Thus the kinetic behaviour is due to the possibility of energy interchange of two kinds, one being with the sources of the field in the form of the original charged conductors defining the system and the other being with the environment in the form of thin conducting sheets.

The explanation of equations (5.9) and (5.11) follows in the same manner by comparison with equations (4.82) and (4.74). In these variational statements we have replaced ϕ by $\mathbf{D}$ as the varied co-ordinate. It should be noted that the choice of potential or kinetic energy does not depend on the choice of co-ordinate but on whether the sources are allowed to vary. In the potential energy formulation the energy appears purely as a field quantity. The sources of the field are implicit. However, in the kinetic energy formulation the sources are stated explicitly.

The notions of potential and kinetic energy have served us well but they are artificial in a system which is independent of time. Instead we can adopt a more general classification and distinguish between convex and concave energy functionals. Convex functionals have a minimum energy at equilibrium and correspond to potential energy, whereas concave functionals like kinetic energy have a maximum energy at equilibrium. Following Noble and Sewell† we shall designate the convex functionals by the letter Y and the

† B. Noble and M. J. Sewell. On dual extremum principles in applied mathematics, *J. Inst. Math. Appl.* **9**, 123–193 (1972).

concave functionals by the letter Z. These functionals occur in pairs and are connected by a double Legendre transformation. Both Y and Z are Lagrangian functionals.

A single Legendre transformation converts either Lagrangian into a Hamiltonian functional which we discussed in § 2.4. As expected the Hamiltonian, which Noble and Sewell designate by the letter X, is neither convex nor concave but saddle shaped. It can be treated as the generating functional of Y and Z.

5.3. The role of the Lagrangian multiplier in generating convex or concave functionals

The Lagrangian scheme is based on the variables q and $\dot{q}$ which embody the distinction between potential and kinetic energy in mechanics. The dual Lagrangian uses the transformed variables p and $-\dot{p}$ which invert the distinction between the types of energy. The inversion is equivalent to a reversal of time as shown by the signs of $+\dot{q}$ and $-\dot{p}$ as adjoint quantities. The idea of such a reversal has its uses but is somewhat artificial.

However, in electrostatic systems we have replaced the differentiation with respect to time by the spatial operators gradient and divergence. These operators are adjoint if one of them has a positive sign and the other a negative one. The choice of sign has no particular physical significance because distance in space has no unique direction. For example we can use ϕ, $-\nabla\phi = \mathbf{E}$ and the dual $\mathbf{D}$, $\nabla \cdot \mathbf{D} = \rho$, or equally well we can use ϕ, $-\mathbf{E}$, $-\mathbf{D}$, ρ. Other schemes are possible, but these two use a positive sign for ϕ and ρ which are the source quantities defining the system.

Let us now consider an electrostatic system in which ϕ varies and ρ is held fixed. We can write the energy functional as

$$W = \tfrac{1}{2}\langle \phi, \rho \rangle + \tfrac{1}{2}[\phi, \sigma] + \langle \lambda(\nabla\phi + \mathbf{E}) \rangle \qquad (5.12)$$

where λ is a Lagrangian multiplier and where we have used W because we do not yet know whether we have a convex or a concave functional. Hence

$$W = \tfrac{1}{2}\langle \phi, \rho \rangle + \tfrac{1}{2}[\phi, \sigma] - \langle \phi, \nabla \cdot \lambda \rangle + [\phi, \lambda_n] + \langle \lambda, \mathbf{E} \rangle. \qquad (5.13)$$

If we put $\lambda = \mathbf{D}/2$

$$W = Y = \tfrac{1}{2}\varepsilon\langle E^2 \rangle \qquad (5.14)$$

and if we put $\lambda = -\mathbf{D}/2$

$$W = Z = \langle \phi, \rho \rangle + [\phi, \sigma] - \tfrac{1}{2}\varepsilon\langle E^2 \rangle. \qquad (5.15)$$

Thus the choice of the sign of λ determines the type of energy which is being considered. The same is true for the alternative variation of ρ and σ. We have

$$W = \tfrac{1}{2}\langle \phi, \rho \rangle + \tfrac{1}{2}[\phi, \sigma] + \langle \lambda, (\nabla \cdot \mathbf{D} - \rho) \rangle \qquad (5.16)$$

which leads to

$$W = \tfrac{1}{2}\langle \phi, \rho \rangle + \tfrac{1}{2}[\phi, \sigma] - \langle \mathbf{D}, \nabla \lambda \rangle + [D_{\mathrm{n}}, \lambda] - \langle \lambda, \rho \rangle. \qquad (5.17)$$

The choice of $\lambda = \phi/2$ leads to

$$W = Y = \frac{1}{2\varepsilon} \langle D^2 \rangle \qquad (5.18)$$

whereas the choice of $\lambda = -\phi/2$ leads to

$$W = Z = \langle \phi, \rho \rangle + [\phi, \sigma] - \frac{1}{2\varepsilon} \langle D^2 \rangle. \qquad (5.19)$$

Underlying these two formulations of the energy are the variables $\mathbf{D}$, ρ, ϕ, $\mathbf{E}$ and $\mathbf{D}$, $-\rho$, $-\phi$, $\mathbf{E}$. In both variations we can therefore use the Lagrangian multiplier to give the type of energy which we wish to consider.

Further light is thrown on the behaviour of the Lagrangian multiplier by the following considerations. As we found in §2.2 the physical significance of the multipliers is that they provide the reaction forces imposed by the kinematical constraints which act in mechanical systems. Consideration of equations (5.13) and (5.17) shows that in our case the multipliers provide the 'reaction sources' which are needed to balance the assigned sources. By a correct choice of the multipliers we can therefore cancel the contribution of these sources to the variation of energy. This converts the problem to a 'free' problem involving potential energy only and giving a convex functional. The choice of a multiplier of opposite sign forces the problem into one having two types of energy, which is of course the characteristic of kinetic energy, which results in a concave functional.

5.4. Energy of static systems in terms of circuit parameters

It is customary and helpful to describe the energy of systems in terms of lumped parameters, which because of their use in circuit and network problems have become known as circuit parameters. This well-established usage is slightly unfortunate because it suggests a restricted sphere of application for such parameters. Their underlying meaning is connected with energy and they are just as applicable to distributed systems as to the highly simplified one-dimensional components of circuits.

The application of Thomson's theorem to the calculation of energy parameters is due to Maxwell who states that he derived the idea from a paper by Lord Rayleigh entitled *On the theory of resonance*.[†] Rayleigh used Thomson's theorem to find approximate values for the frequency of vibrations in organ pipes. Maxwell developed the method to calculate capacitance and resistance.[‡]

[†] Lord Rayleigh, On the theory of resonance, *Phil. Trans. R. Soc.* **161**, 77–118 (1871).
[‡] J. C. Maxwell, *Electricity and magnetism*, Articles 102 and 306.

Let us consider the calculation of capacitance. In § 5.2 we have four variational principles for the electrostatic energy. How can these be used to find capacitances? In equation (5.8) the charges are specified and the potential is varied. In terms of capacitance we have

$$\delta\left(\frac{\bar{Q}^2}{2C}\right) = 0. \tag{5.20}$$

Since the functional in equation (5.8) is concave, the variational statement implies a maximum value for energy and a minimum value for the capacitance at equilibrium. Hence any approximate value will provide an upper bound for the capacitance which we shall denote by C_+.

Equation (5.9) can be recast into the form

$$\delta(\tfrac{1}{2}\bar{\phi}^2 C) = 0. \tag{5.21}$$

It is also a concave functional and thus provides a lower bound for the capacitance which we write C_-. This lower bound is associated with a variation of the flux density **D**.

In equation (5.10) we have a variation of ϕ, but only in the space between the sources. The sources are implicit in the functional in terms of **E**. Therefore the problem is specified in ϕ and equation (5.21) applies. However, the functional is convex and therefore provides an upper bound C_+. Similarly equation (5.11) provides a lower bound C_-.

We notice that the four equations provide two possibilities for an upper-bound value of the capacitance and two for a lower bound. The type of bound depends on the choice of variables and the two possibilities for each bound arise because the problem can be specified either in charges or potentials. It is of course reasonable that the capacitance bound is determined only by the variation and not by the specification of the type of source. The reader will remember that the variation involves the insertion of conducting sheets or of flux barriers. The former increase the capacitance and the latter reduce it.

There is little difficulty in applying the above analysis to resistive systems carrying steady current and to magnetostatic systems.

For resistive systems we refer to § 3.3.3. In that section we dealt with power rather than energy, but since the power is constant in time the energy could be obtained by multiplying throughout by a constant time factor. Equations (3.73) and (3.76) give the two concave functionals. The former has a variation of ϕ and gives a lower bound R_- for the resistance. The latter gives R_+ by varying the current. The two convex functionals can easily be derived. They are

$$Y = \frac{1}{2\sigma}\langle E^2 \rangle \tag{5.22}$$

and

$$Y' = \frac{\sigma}{2} \langle J^2 \rangle. \tag{5.23}$$

$\delta(Y)$ gives R_- and $\delta(Y')$ gives R_+. The physical significance of the variation is particularly clear because the insertion of highly conducting sheets must lower the resistance whereas the insertion of insulation sheets raises the resistance. It should be noticed that the conductance, which is the reciprocal of the resistance, has the same type of behaviour as the capacitance.

Magnetostatic systems were discussed in § 3.3.2. The convex functionals are given in equations (3.63) and (3.64). The concave functionals can be simplified either in terms of magnetic poles as in equations (3.53) and (3.54) or in terms of electric current as in equations (3.60) and (3.62). In all these cases the variation in **H** gives a higher value of the inductance L_+ and the variation in **B** gives the lower value L_-.

5.5. Numerical examples of the calculation of resistance parameters in static systems

The object of this section is the acquisition of experience of the variational method. We shall examine very simple problems and problems which can be solved by other means so that we can test the accuracy of our method against known solutions. Our ultimate aim is the solution of complicated problems, but we shall proceed cautiously.

5.5.1. A simple resistance network

Consider the network of Fig. 5.3. It is required to find the input resistance at the terminals AB. A simple series/parallel calculation gives $0.625R$. We now use the variational method. This is based on Tellegen's theorem (equation (3.22)) which depends on the independence of Kirchhoff's current and voltage laws. Consider first a set of currents obeying the node law but giving wrong voltages as in Fig. 5.4. The power is $(1^2R + 2^2R + 1^2R + 1^22R) = 7R$. The input current is 3, so the input resistance is $\frac{7}{9}R = 0.778R$. Now

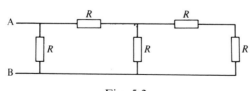

Fig. 5.3

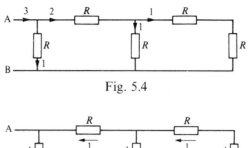

Fig. 5.4

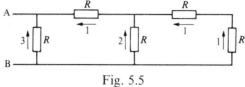

Fig. 5.5

consider a voltage system obeying the loop equations as in Fig. 5.5. The power is

$$\frac{9}{R} + \frac{1}{R} + \frac{4}{R} + \frac{1}{R} + \frac{1}{R} = \frac{16}{R},$$

The input voltage is 3 and hence the input resistance

$$\frac{9}{16}R = 0.563R.$$

The average of the upper and the lower bounds is $0.670R$, giving an error of 7 per cent. This might be thought a very close estimate in view of the rather wild choices of the currents and voltages. We note that R_+ is given by altering the flow of current and R_- by altering the potentials. This is in accordance with the discussion in § 5.4.

5.5.2. Resistance of a trapezoidal conducting plate

Consider the plate shown in Fig. 5.6. It is required to find the resistance between the vertical sides. The conductivity is σ and the thickness of the plate is b.

Consider first a possible current flow as shown in Fig. 5.7. The typical strip has a resistance

$$\delta R = \frac{1}{\sigma b} \frac{2 \sec \theta}{1.5 \, \delta y \cos \theta}$$

and $y = 2 \tan \theta$. Hence the admittance is given by

$$\frac{1}{R} = \int_0^{\tan^{-1} 1\frac{1}{2}} 1.5 \sigma b \, d\theta = 0.6955 \sigma b$$

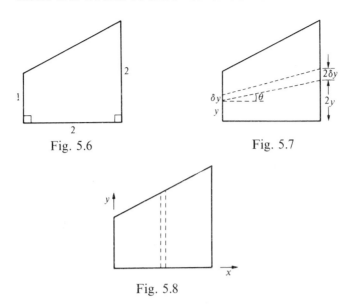

Fig. 5.6   Fig. 5.7

Fig. 5.8

and $R_+ = 1.438/\sigma b$. This value is slightly suspect because we have used the average width of a flow tube in Fig. 5.7. Also the ends of these tubes are incorrectly modelled because the current has to be perpendicular to the vertical sides as it enters and leaves the plate. We shall not pursue these matters further but leave them to the interested reader.

Consider now a possible potential distribution as shown in Fig. 5.8:

$$\delta R = \frac{\delta x}{\sigma b y}$$

where $y = 1 + x/2$. Hence

$$R_- = \frac{2}{\sigma b}\ln 2 = \frac{1.386}{\sigma b}.$$

The average of R_+ and R_- is $1.412/\sigma b$ and the difference is 3.65 per cent of the average value. We can assert with some confidence that the average value is within 2 per cent of the exact value.

The purpose of this discussion is to show that the variational method provides an extremely simple way of achieving an accuracy which is sufficient for most purposes. An ordinary field solution would require a very much greater computational effort. The chief difference is that the variational method does not attempt to find the correct angles between the flow lines and the equipotentials, but we have shown that these angles do not greatly affect the resistance parameter. In other words we do not need a solution of Laplace's equation to find the resistance.

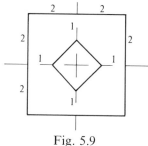

Fig. 5.9

5.5.3. Resistance of a plate with a central hole

Figure 5.9 illustrates the problem. We seek an approximate value of the resistance between the vertical sides of the plate. To anchor our discussion a finite-difference solution was obtained which gave $R = 1.34/\sigma b$. Because of symmetry we need consider only a quarter of the plate. Figure 5.10 shows a possible flow tube. If θ is the angle between the tube and the horizontal, the length of the typical tube is $2 \sec \theta$ and the average width is $1.5 \, \delta y \cos \theta$. Also $\tan \theta = y/2$. Hence the admittance is

$$\frac{1}{R} = \sigma b \int_{0}^{\tan^{-1}\frac{1}{2}} \tfrac{3}{2} \, d\theta = 0.695 \sigma b$$

and

$$R_{+} = \frac{1.44}{\sigma b}$$

A lower bound can be obtained as illustrated in Fig. 5.11:

$$R_{-} = \frac{1}{\sigma b} \left[\int_{0}^{1} \frac{dx}{y} + \int_{1}^{2} \frac{dx}{2} \right]$$

$$= \frac{1}{\sigma b} (\ln 2 + 0.5) = \frac{1.19}{\sigma b}.$$

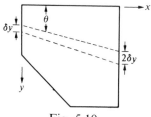

Fig. 5.10

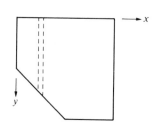

Fig. 5.11

The average of our bounded values is $R = 1.315/\sigma b$ which is within 2 per cent of the accurate solution.

The bounded values in this example and in the previous one have been obtained by the use of thin insulating and conducting sheets. We discussed this variation method in the context of Thomson's theorem in § 4.6. The method guarantees that the bounded values vary as the second order of the field variation and this produces values close to the equilibrium energy.

The method requires geometrical constructions or sketches and is particularly suited to the design office. Flux maps have of course been used for many years. The variational method gives the additional insight that double-bounded solutions are obtainable by sketching both the flux distribution and the potential distribution. This means that in two-dimensional fields it is not necessary to attempt a single consistent map of flux and potential lines in the form of curvilinear squares. It also makes possible the analysis of three-dimensional fields, to which the method of curvilinear squares does not apply. The geometrical method has therefore very much to recommend it.

However, in computational work geometrical methods are inconvenient because they may require somewhat complicated computer instructions. An alternative algebraic formulation may then be preferable. Let us look at this possibility in the context of the problem of the resistance of the plate of Fig. 5.9.

Consider first the variation of the potential which gives a lower bound for the resistance.

We make use of the concave functional of equation (3.76):

$$Z = -b[\bar{J}_n, \phi] - \frac{\sigma b}{2} \langle E^2 \rangle. \tag{5.24}$$

Suppose we model the potential as varying linearly between the vertical edges of the plate. For the quarter plate of Fig. 5.11 we put $\phi = \alpha - \beta x$ where α and β are unknown coefficients. We can choose a current of unity which gives

$$Z = \alpha - \alpha + 2\beta - \frac{\sigma b}{2} 3.5\beta^2 \tag{5.25}$$

where 3.5 is the area of the plate and $E = \beta$. We now optimize Z by putting $\partial Z/\partial \beta = 0$. This gives

$$\beta = \frac{2}{3.5\sigma b} \quad \text{and} \quad Z = \frac{2}{3.5\sigma b}.$$

Also $Z = \frac{1}{2}I^2R$, whence $R_- = 1.14/\sigma b$. We can seek to improve the value of R by using a higher-order polynomial for the potential. Consider $\phi = \alpha - \beta x - \gamma x^2$:

$$Z = 2\beta + 4\gamma - \frac{\sigma b}{2} \int_0^1 (\beta + 2\gamma x)^2 (1 + x)\, dx$$

$$- \frac{\sigma b}{2} \int_1^2 (\beta + 2\gamma x)^2 2\, dx$$

$$= 2\beta + 4\gamma - \frac{\sigma b}{2}(3.5\beta^2 + 15.34\beta\gamma + 21\gamma^2). \qquad (5.26)$$

To optimize we put $\partial Z/\partial\beta = \partial Z/\partial\gamma = 0$, which gives $2 - \sigma b(3.5\beta + 7.67\gamma) = 0$ and $4 - \sigma b(7.67\beta + 21\gamma) = 0$. Hence $\beta = 0.772/\sigma b$, $\gamma = -0.086/\sigma b$, $Z = 0.6/\sigma b$, and $R_- = 1.2/\sigma b$. Thus the lower bound of the resistance has been improved by 4 per cent.

Consider now the upper bound. The Z functional is given in equation (3.79) as

$$Z = -b[\bar{\phi}, J_n] - \frac{b}{2\sigma}\langle J^2 \rangle. \qquad (5.27)$$

The variation is defined in § 4.6 where we found that the divergence of the flux or current can be varied only where the potentials are fixed. The reason for this is that a variation of charge alters the potentials unless these have assigned values. This imposes a restriction because our choice of J must not have any divergence in the volume of the conductor nor must it have a normal component except at the electrodes. This forces us either to return to the geometrical specification or towards an iterative procedure which makes use of the dual specification. The choice will depend on the problem to be solved and on the experience of the analyst. The example of the plate with the central hole could easily be cast into algebraic form using both J_x and J_y as functions of the angle θ as shown in Fig. 5.10.

An iterative procedure could be devised as follows. Consider the specification $J_x = \alpha$, $J_y = 0$. This gives

$$Z = b\alpha - \frac{3.5}{2\sigma} b\alpha^2. \qquad (5.28)$$

Taking $\partial Z/\partial\alpha = 0$ gives $\alpha = \sigma/3.5$ and $Z = \sigma b/7$. This gives $R_+ = 3.5/\sigma b$ which is very much too high. We notice that there are no divergence sources in the volume, but at the sloping side there is a faulty $J_n = \alpha/\sqrt{2}$. If we choose the potential ϕ to be decreasing linearly with x there is a faulty energy

$$b[\phi, J_n] = b_4^3 \frac{\alpha}{2} \sqrt{2} = 0.53b,$$

which should be added to the surface term of Z and which will give a revised value $R_+ = 1.5/\sigma b$, which is of course very much closer to the accurate value. We shall not pursue this matter further but we hope that the discussion of the example in this section has shown that a variety of techniques is available to give an accuracy within a few per cent without requiring more than a few minutes of calculation using a simple desk calculator.

5.6. Optimization of the Z functional

In the example of § 5.5.3 we have used an optimization procedure for the Z functional rather than for the Y functional and the reason for this needs to be briefly stated.

Consider the electrostatic case

$$Z = [\phi, \bar{\sigma}] + \langle \phi, \bar{\rho} \rangle - \frac{\varepsilon}{2} \langle E^2 \rangle. \tag{5.29}$$

The optimization procedure equalizes the surface and volume energies so that

$$[\phi, \bar{\sigma}] + \langle \phi, \bar{\rho} \rangle = \varepsilon \langle E^2 \rangle. \tag{5.30}$$

However, we also know from Green's first identity (equation (4.2)) that

$$\left[\phi, \frac{\partial \phi}{\partial n} \right] - \langle \phi, \nabla^2 \phi \rangle = \varepsilon \langle [\nabla \phi]^2 \rangle \tag{5.31}$$

which we can write as

$$[\phi, \sigma] + \langle \phi, \rho \rangle = \varepsilon \langle E^2 \rangle. \tag{5.32}$$

Hence by comparison of equations (5.30) and (5.32) we see that the optimization draws the energy of the varied sources towards the energy of the assigned sources interacting with the varied potentials.

In the Y functional we start with

$$Y = \frac{\varepsilon}{2} \langle E^2 \rangle. \tag{5.33}$$

Using equation (5.32) we have

$$Y = \tfrac{1}{2}[\phi, \sigma] + \tfrac{1}{2}\langle \phi, \rho \rangle. \tag{5.34}$$

Moreover, the surface condition of $\bar{\sigma} = \varepsilon(\partial\phi/\partial n)$ is introduced implicitly by the specification of E at the surface. However, the volume sources are not specified. All we can state is

$$[\phi, \bar{\sigma}] + \langle\phi, \rho\rangle = \varepsilon\langle E^2\rangle. \tag{5.35}$$

Therefore the optimization procedure does not use the volume sources and will not in general converge without an iterative procedure. Whereas the accuracy of the system parameters obtained from a Z functional can be improved by resorting to higher-order polynomials, this is not true for the Y functionals.

5.7. Calculation of capacitance

The fundamental problem in calculating capacitance is that the charge distribution is generally unknown whereas the potentials are fixed. Consider the problem illustrated by Fig. 5.12 which shows a tubular capacitor of square section.† If the section were circular the charge distribution would be

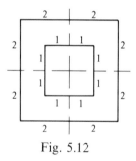

Fig. 5.12

uniform because of symmetry, but the corners of the square section introduce a difficulty. However, there is sufficient symmetry for us to consider only one section as shown by Fig. 5.13.

For the potential variation we can use vertical strips as shown in Fig. 5.13:

$$\delta C = \varepsilon y/\delta x \quad \text{per unit length.}$$

Hence

$$\frac{1}{\delta c} = \frac{\delta x}{\varepsilon(1 + x)}$$

---

† The examples of §§ 5.7, 5.8, and 5.9 are also discussed in P. Hammond and J. Penman, Calculation of inductance and capacitance by means of dual energy principles, *Proc. Inst. Electr. Eng.*, **123** (6), 554–559.

and

$$\frac{1}{C} = \frac{1}{\varepsilon} \ln 2 = \frac{0.693}{\varepsilon},$$

This gives $C_+ = 1.44\varepsilon$ for the part of the capacitor shown in Fig. 5.13. Hence for the whole tube $C_+ = 11.52\varepsilon$.

The lower bound is easily obtained by drawing flux lines as shown in Fig. 5.14:

$$\delta C = \varepsilon \times 1.5 \cos \theta \, \delta y/\sec \theta \quad \text{where } \tan \theta = y$$
$$\delta C = \varepsilon \times 1.5 \, \delta\theta$$
$$C_- = 8 \times 1.5 \times \pi/4\varepsilon = 9.42\varepsilon.$$

The average of C_+ and C_- is $C = 10.48\varepsilon$. An analytical solution by means of a conformal transformation gives $C = 10.25\varepsilon$ so that the variational method comes within 2.3 per cent of the exact solution.

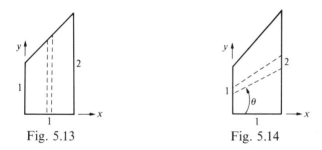

Fig. 5.13   Fig. 5.14

Although it is of course possible to improve the approximation this will often be unnecessary. The great advantage of the double-bounded solution is that the average of the bounds is very stable. In this particular example the upper bound is within 12.6 per cent of the exact value and the lower bound within 9.9 per cent. Neither of these values is very satisfactory, but the average value is close enough for most purposes. The strength of the method is that relatively rough guesses give a good working answer.

5.8. Calculation of inductance

Ohm's law states that the current density in ordinary parallel-sided conductors is uniform provided that the frequency is low enough to ensure that there is a negligible skin effect. Exact values of inductance can be calculated for such conductors because the sources of the magnetic field are known. However, if ferromagnetic materials are included in the system, there are also unknown sources due to the magnetic pole strength. In such cases the variational method is particularly helpful.

Consider the internal inductance of the conductor shown in Fig. 5.15 which is surrounded by highly permeable iron except on its top surface. An exact answer has been obtained for this inductance[†] and its value is $L = 0.57\mu_0$ per unit length. This is the value with which we shall compare the approximate solutions obtained by the energy method. Let there be a uniform current density of strength unity in the conductor. Since the area is 10, the current is also 10.

Consider first the lower bound of the inductance which can be obtained by using suitable flux barriers as shown in Fig. 5.16. The flux density is forced to have only an x component. We assume zero flux density at the bottom of the conductor because no current is enclosed. Consider a flux barrier just below the shoulders of the conductor and another one at the bottom of the narrow portion. These barriers split the region into two.

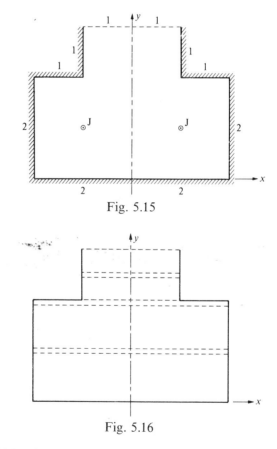

Fig. 5.15

Fig. 5.16

† D. E. Jones, N. Mullineux, J. R. Reed, and R. L. Stoll, Solid rectangular and T-shaped conductors in semi-closed slots, *J. Eng. Math.*, **3**, 123–135 (1969).

The energy is given by

$$\tfrac{1}{2}\mu_0 \int_0^2 4B^2 \, dy + \tfrac{1}{2}\mu_0 \int_2^3 2B^2 \, dy.$$

In the bottom section

$$B = \mu_0 H = -\mu_0 y$$

and in the top section

$$B = \mu_0 H = -4\mu_0 - (y - 2)\mu_0 = -(2 + y)\mu_0.$$

Since the current is given and the flux is varied, we are dealing with a Z functional:

$$Z = \frac{2}{\mu_0} \int_0^2 \mu_0 y^2 \, dy + \frac{1}{\mu_0} \int_2^3 (2 + y)^2 \mu_0^2 \, dy.$$

Hence

$$Z = 25.67\mu_0$$

and

$$L_- = 0.513\mu_0 \quad \text{per unit length.}$$

To obtain an upper bound we must insert highly permeable sheets into the system. Consider this with reference to Fig. 5.17. Two typical sheets are shown; the spacing between them is δx at the top of the conductor and $2\,\delta x$ at the bottom. The sheets are parallel in the top portion and diverge in the

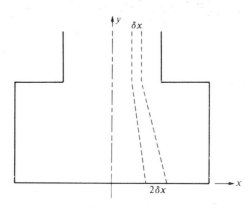

Fig. 5.17

bottom portion. The angle between the sloping side and the vertical is θ. We assume that the flux density varies linearly from the bottom to the top:

$$\delta L = \tfrac{1}{2}\mu_0 \left(\frac{2}{1.5\,\delta x \cos^2\theta} + \frac{1}{\delta x} \right)$$

$$\frac{1}{\delta L} = \frac{2}{\mu_0} \frac{\delta x}{1 + 1.33\sec^2\theta} = \frac{8}{\mu_0} \frac{\delta\theta}{3.67 + \cos 2\theta}$$

and θ varies from $\tan^{-1}\tfrac{1}{2}$ to $\tan^{-1}(-\tfrac{1}{2})$. Using a standard integral† we obtain

$$\frac{1}{L} = \frac{8}{\mu_0} \frac{1}{(3.67^2 - 1)^{1/2}} \left[\tan^{-1}\left(\frac{2.67}{4.67}\tan\theta \right) \right]_{-26.56}^{+26.56}$$

and $L_+ = 0.613\mu_0$ per unit length. The average inductance is $L = 0.563\mu_0$ which is within 1.23 per cent of the exact value.

For an algebraic method we can employ the Z functional

$$Z = [\mathbf{A} \times \bar{\mathbf{H}}, \hat{\mathbf{n}}] + \langle \mathbf{A}, \bar{\mathbf{J}} \rangle - \frac{1}{2\mu} \langle B^2 \rangle. \tag{5.36}$$

At the top of the conductor $\bar{H} = -5\bar{J}$ and, since $\mathbf{H}$ is zero along the iron boundaries,

$$Z = -[\mathbf{A}, 5\mathbf{J}] + \langle \mathbf{A}, \bar{\mathbf{J}} \rangle - \frac{1}{2\mu} \langle B^2 \rangle. \tag{5.37}$$

A trial function of $\mathbf{A}$ which gives a linear variation of $\mathbf{B}$ is $A = \alpha y + \beta y^2$, which on substitution into equation (5.37) gives

$$Z = (-17\alpha - 66.67\beta)J - \frac{1}{2\mu}(10\alpha^2 + 52\alpha\beta + 93.33\beta^2). \tag{5.38}$$

Putting

$$\frac{\partial Z}{\partial \alpha} = \frac{\partial Z}{\partial \beta} = 0$$

we obtain $\alpha = 0.56\mu_0$, $\beta = -0.87\mu_0$, and $Z = 24.2J\mu_0$. Since the current is $10J$, this gives $L_- = 0.485\mu_0$ per unit length. This is somewhat worse than the previous value but it avoids the division of the conductor into two regions.

It is more difficult to obtain L_+ because H must ideally be correct at the boundaries and in the volume. A rough simple choice is to put $H_x = -\tfrac{5}{3}yJ$, $H_y = 0$ which satisfies the bottom, side and top surfaces. It does not satisfy

† Integral 14.390 in M. R. Spiegel, *Mathematical Handbook*, McGraw-Hill, New York, 1968.

the shoulders nor the volume current density:

$$Y = \tfrac{1}{2}\mu_0 \left(\int_0^2 4H^2 \, dy + \int_2^3 2H^2 \, dy \right) = \tfrac{1}{2}\mu_0 \times 64.82. \qquad (5.39)$$

Hence $L_+ = 0.648\mu_0$.

The average value is $L = 0.567\mu_0$ which is within an accuracy of just over 0.5 per cent of the exact value. Both the geometrical and the algebraic methods are very accurate.

5.9. A non-linear problem

We have already mentioned in §§ 2.6 and 3.2 that the energy method is applicable to non-linear relationships. In a magnetic problem, for example, one solution is obtained by varying B and the dual by varying H. The relationship between B and H does not enter into the variational process as long as the energy relationships $\int H \, dB$ and $\int B \, dH$ can be uniquely defined.

As an example consider the internal energy of a conductor of circular cross-section of radius a. Let the B–H curve be nearly rectangular as shown in Fig. 5.18. We can put $\int H \, dB = 0$ and $\int B \, dH = B_s H$. Let the current density be J; then $H = JR/2$ where R is the radius and $H = H_\phi$ is circumferential in direction. Consider the Z functional and choose $A_3 = -B_s R$ so that $(\text{curl } A)_\phi = B_s$:

$$Z = +B_s J a^3 \pi - \tfrac{2}{3} B_s J a^3 \pi$$
$$= \tfrac{1}{3} a^3 \pi B_s J = 1.05 a^3 B_s J.$$

Consider next the Y functional:

$$Y = \left\langle \int_0^H B \, dH \right\rangle = \langle B_s, JR/2 \rangle$$

$$= \tfrac{1}{3} a^3 \pi B_s J = 1.05 a^3 B_s J.$$

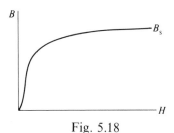

Fig. 5.18

Therefore in this idealized case of a fully rectangular B–H curve the two functionals are equal and give the exact value of energy.

Consider next a square conductor. This has less symmetry than a circular conductor and it is not possible to give a simple description of the field. A lower bound on the energy can be obtained by using circular flux barriers and terminating the fields on the barrier which has a radius of a. This gives the same result as the Z energy for a circular conductor. Thus

$$Z = \frac{\pi}{3} a^3 B_s J = 1.05 a^3 B_s J.$$

For the Y functional we can use the energy inside the circular region and add the energy in the corners as shown in Fig. 5.19. The area in the corners is $(4 - \pi)a^2$ and we can take an average $H = (\sqrt{2} - 1)a(J/2)$. Then

$$Y = \frac{\pi}{3} a^3 B_s J + (4 - \pi)(\sqrt{2} - 1)a^3 \frac{J}{2} B_s = 1.23 a^3 B_s J.$$

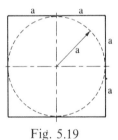

Fig. 5.19

5.10. Resistance and inductance calculations for conductors carrying alternating current

If the current is alternating the resistance and inductance energies are no longer independent of each other. This means that the equilibrium distribution of the current cannot be determined by a simple application of Ohm's law even if the conductors have a constant cross-section. The variational method is therefore particularly appropriate because it provides a means of finding the energy parameters without having to determine the exact distribution of the current or of the magnetic field.

We have already discussed the principles of the method in § 3.3.4 where we faced the difficulty that the diffusion equation (equation (3.84)) contains a $\partial/\partial t$ term and is therefore not self-adjoint. Many writers have decided that this precludes the use of the variational method. Physically the $\partial/\partial t$ term represents the conversion of electromagnetic energy to heat energy which is lost to the system. Energy is therefore not conserved in the system and the idea of equilibrium of the system under small displacements seems to be inapplicable. We note in passing that the difficulty is caused because the system is incompletely defined if thermodynamic processes are excluded. However, we do not wish to introduce such complications into the calculation of resistance and inductance which we regard as purely electromagnetic quantities.

We have seen that the difficulty can be removed most easily by using the idea of energy sources which make up the loss of energy due to heat dissipation. Since we are dealing with a loss mechanism distributed in space the sources must have an identical distribution. Since the dissipation also varies in time the sources must also show such a variation. This means that the energy balance must be correct at each instant and we have to restrict the variational processes to deal with rate of energy, or power, rather than with energy.

The power sources can be dealt with by postulating an adjoint system so that the original system together with the adjoint system can be treated as an isolated self-adjoint system. If the time variation is harmonic and is expressed in the phasor notation of complex numbers the adjoint field quantities are the complex conjugates of the original quantities. Two of the variational principles were given in equations (3.93) and (3.98).

The four power functionals are

$$Z = -[\mathbf{E} \times \bar{\mathbf{H}}^*, \hat{\mathbf{n}}] - \frac{\sigma}{2} \langle \mathbf{E}^*, \mathbf{E} \rangle - \frac{j\omega}{2\mu} \langle \mathbf{B}, \mathbf{B}^* \rangle \qquad (5.40)$$

$$Y = +\frac{\sigma}{2} \langle \mathbf{E}, \mathbf{E}^* \rangle + \frac{j\omega}{\mu} \langle \mathbf{B}, \mathbf{B}^* \rangle \qquad (5.41)$$

$$Z = -[\bar{\mathbf{E}} \times \mathbf{H}^*, \hat{\mathbf{n}}] - \frac{1}{2\sigma} \langle \mathbf{J}, \mathbf{J}^* \rangle - \frac{j\omega\mu}{2} \langle \mathbf{H}, \mathbf{H}^* \rangle \qquad (5.42)$$

$$Y = +\frac{1}{2\sigma} \langle \mathbf{J}, \mathbf{J}^* \rangle + \frac{j\omega\mu}{2} \langle \mathbf{H}, \mathbf{H}^* \rangle. \qquad (5.43)$$

The asterisk denotes the complex conjugate. It is important to note that the functionals are specified in terms of the double system, which would of course have been expected. Consider, for example, the variation of $\langle \mathbf{J}, \mathbf{J}^* \rangle$. Suppose that $J_x = J_x' + jJ_x''$ is complex. This means that J_x, $J_x^* = J_x'^2 + J_x''^2$. Hence the variation acts only on the magnitude and not on the phase angle of the field quantity. The variation method restricts the information to be handled by ignoring local phase angles, and in this lies its great advantage. The chief difficulty in eddy current calculations arises from the fact that a complete field solution requires both a local magnitude and a local phase angle. The variational method calculates only a system magnitude and a system phase angle. Since the information about local phase angles is hardly ever of interest in eddy current problems, the variational method is the sensible choice.

The four functionals of equations (5.40)–(5.43) give dual bounds on the resistance and inductance as follows.

If the system is specified in terms of total current the power is given by

$$P = \tfrac{1}{2}I^2(R + jX).\tag{5.44}$$

The variation of $\mathbf{E}$ and $\mathbf{B}$ gives R_+ and X_- and the variation of $\mathbf{H}^*$ and $\mathbf{J}^*$ gives R_- and X_+. If the system is specified in terms of total flux and hence of applied voltage

$$P = \tfrac{1}{2}V^2\left(\frac{1}{r} + j\,\frac{1}{x}\right).\tag{5.45}$$

The variation of $\mathbf{E}$ and $\mathbf{B}$ gives r_+ and x_- and the variation of $\mathbf{H}^*$ and $\mathbf{J}^*$ gives r_- and x_+.

At first sight it seems strange that the bounds of resistance and inductance have opposite signs. This arises from the use of the adjoint system. Thus if the current is specified we have to vary the electric field of the adjoint system which has negative conductivity. Similarly a specification of the voltage requires a variation of the adjoint current system. Alternatively the adjoint current and actual voltage can be specified.

5.10.1. Resistance and inductance of a large flat conductor†

The system is illustrated in Fig. 5.20. We discuss this problem because it has an analytical solution given by

$$R + jX = \frac{1}{2\sigma\Delta}\left(\frac{\sinh 2b/\Delta + \sin 2b/\Delta}{\cosh 2b/\Delta - \cos 2b/\Delta} + j\,\frac{\sinh 2b/\Delta - \sin 2b/\Delta}{\cosh 2b/\Delta - \cos 2b/\Delta}\right)\tag{5.46}$$

per unit in the x and z directions, where $\Delta = \sqrt{2/\mu\sigma\omega}$ is the eddy-current skin depth.

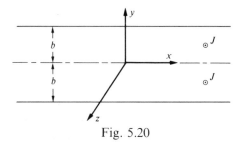

Fig. 5.20

† The results in this and the following section are taken from P. Hammond and J. Penman, Calculation of eddy currents by dual energy methods, *Proc. Inst. Electr. Eng.* **125** (7), 701–708 (1978).

Two types of behaviour are expected. If $b > \Delta$ the conductor is said to be thick, and if $b < \Delta$ it is said to be thin. For a thick conductor there is little interaction between the surfaces and each surface layer can be treated independently. For a thin conductor there is considerable interaction between the surfaces. We shall use a different approximation for the two cases.

Consider first a thick conductor with $b > 2\Delta$. We assume a surface layer which penetrates only to the depth of 2Δ. Suppose we use the bottom layer and fix the origin of the y co-ordinate on the surface. Let there be a total current I in the conductor. Then by symmetry the magnetic field is $H_x = I/2$ at the bottom surface. Let the magnitude of the in-phase and quadrature components of H_x be given by

$$H' = I/2 + \alpha_1 y + \alpha_1{}^2$$
$$H'' = \beta_1 y + \beta_2 y^2 \tag{5.47}$$

Let $H' = H'' = 0$ at $y = 2\Delta$.

The current density is given by $\mathbf{J} = \text{curl } \mathbf{H}$:

$$J' = -\alpha_1 - 2\alpha_2 y$$
$$J'' = -\beta_1 - 2\beta_2 y. \tag{5.48}$$

Substitute into equation (5.43) and put the variation with respect to the coefficients equal to zero. This gives

$$Y = \frac{1}{2}\frac{I^2}{\sigma\Delta}(0.453 + j0.526). \tag{5.49}$$

For large values of $2b/\Delta$ the analytical solution gives

$$R + jX = (0.5 + j0.5)/\sigma\Delta. \tag{5.50}$$

For the dual bounds, using a quadratic polynomial in E we obtain from equation (5.40)

$$Z = \frac{1}{2}\frac{I^2}{\sigma\Delta}(0.549 + j0.493). \tag{5.51}$$

The average values of R and X give R as 0.2 per cent high and X as 1.9 per cent high compared with the analytical solution. These close estimates have been obtained without the use of hyperbolic and circular functions.

For a thin conductor we have to take account of both surfaces of the conductor. We take the origin of co-ordinates at the centre plane and choose odd functions of y for H and even functions for E. The symmetry conditions for the field and the energy require at least two independent coefficients. Thus we choose

$$H' = \alpha_1 y + \alpha_3 y^3 + \alpha_5 y^5$$
$$H'' = \beta_1 y + \beta_3 y^3 + \beta_5 y^5 \qquad (5.52)$$

subject to $H' = -I/2$ for $y = b$.

It is convenient to use the non-dimensional parameter b/Δ in the solution. The functionals are found to consist of terms of order $(b/\Delta)^{4n}$ of alternating sign. The results shown in Table 5.1 were obtained by neglecting terms of the order $(b/\Delta)^{12}$. The resistances have been normalized to the d.c. resistance $R_0 = 1/2\sigma b$ and the inductances to the d.c. inductance $L_0 = \mu_0 b/6$.

TABLE 5.1

b/Δ	R_-/R_0	R_+/R_0	R/R_0 (analytical)	L_+/L_0	L_-/L_0	L/L_0 (analytical)
0	1	1	1	1	1	1
1	1.08552	1.08554	1.08564	0.97571	0.97570	0.97559

The average values of both the resitance and the inductance lie within 0.01 per cent and the analytical value. It will be noticed that R_+ and L_- lie on the wrong side of the average. This is due to the truncation error which affects the Z functional more than the Y functional. For values of $b/\Delta < 1$ the error disappears and the accuracy increases. It is quite possible to obtain an accuracy of 1 in 10^6, but even for $b/\Delta = 1$ the variational method is remarkably accurate. It can be safely concluded that the variational method is particularly well suited to problems involving the diffusion equation.

The choice of trial functions for H and E implies the insertion of magnetic or electric equipotentials and this similarity of choice makes the Y and Z functionals behave in a similar manner, a behaviour which is reinforced by the fact that the energy diffuses inwards from the surface in the representation of both functionals. In static problems the equipotentials have only a second-order effect on the energy. In eddy current problems the overall effect of the additional energy can similarly be kept small because some of the additional energy is positive and some negative.

5.10.2. A problem having faulty specification of the boundary conditions

The simple geometry of the flat conductor made it easy to match the trial functions to the geometrical boundaries. The purpose of our method, however, is to find approximate solutions to complicated problems. We now discuss such a problem for which an accurate computer solution exists.†

---

† D. E. Jones, N. Mullineux, J. R. Reed and R. L. Stoll, Solid rectangular and T-shaped conductors in semi-closed slots, *J. Eng. Math.*, **3**, 123–135 (1969).

Figure 5.21 shows an inverted T-bar conductor such as is used in induction motor rotors to cause the resistance to vary with frequency. The conductor is surrounded by highly permeable iron except at the top which is open. The boundary conditions are therefore $H_t = 0$ on the iron surface and $H_t = -I/2c$ at the top.

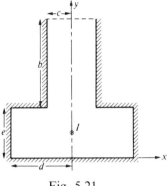

Fig. 5.21

Suppose we assume a trial function for H which gives $H_y = 0$ and $H_x = H' + jH''$ where

$$H' = \alpha_1 y + \alpha_2 y^2$$
$$H'' = \beta_1 y + \beta_2 y^2. \qquad (5.53)$$

These values for H can be matched at the top of the conductor, but there is a difficulty at the junction of the narrow part with the wider part. This difficulty was surmounted by accurate matching in the paper which we are taking as our reference point. In §5.8 we discussed a static problem of similar geometry, but the geometrical device of splitting the region which was described in Fig. 5.16 is not available to us here because we do not know the current distribution. All the energy diffuses into the conductor from the top of the narrow section. Suppose we accept the faulty specification of H on the shoulder and calculate R_- and L_+. If the skin depth Δ is smaller than b the diffusion process will confine the energy largely to the narrow part of the conductor and the faulty specification will not matter. However, if the frequency is reduced so that Δ is greater than b the Y functional will become unreliable. In particular it will be reduced by the power associated with the faulty Poynting vector on the iron shoulder. A similar argument applies to the Z functional in which the surface term will be too large by the amount of power on the shoulder. In fact the faulty power will be transferred from the Y to the Z functional. This, however, need not worry us if we use the average of the values obtained from the Y and Z functionals. The method has the remarkable property of maintaining the average values close to the correct ones. Of course this is not exactly so unless the trial functions of H and E are closely matched.

Figures 5.22 and 5.23 give computed results compared with the exact results. The trial functions for E were quadratic, just as those for H. It will be seen that the bounded solutions for the resistance cross over at a frequency of about 10 Hz and that they are inverted below this frequency.

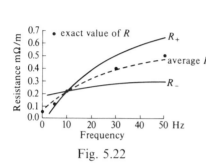

Fig. 5.22

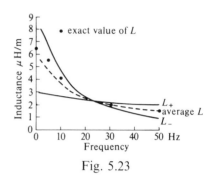

Fig. 5.23

The dimensions in metres for which the computation was made were $c = 0.002$, $b = 0.025$, $e = 0.017$, and $d = 0.0085$ and the conductivity was taken as 0.5×10^8. The skin depth Δ at 10 Hz is 0.0225 and this confirms the discussion about the importance of the shoulders of the conductor. The inductance is affected slightly more and the values cross when $\Delta = 0.015$ which is 60 per cent of the value of b, the length of the narrow portion. The skin depth is defined as reducing the field to 37 per cent of its surface value, so it is not surprising that the values are affected even when Δ is smaller than b.

Although too much importance must not be attached to a single numerical example, the results of Figs. 5.22 and 5.23 increase our confidence in the accuracy of the method of dual bounds for diffusion problems. The transfer of power between the Y and Z functionals causes the average values of the resistance and inductance to be remarkably insensitive to the failure of the trial function in matching the boundary conditions exactly.

5.11. Calculation of radiation resistance

We have already seen in § 3.3.5 that the energy method is applicable to the calculation of electromagnetic wave phenomena. Although a full account of such phenomena is beyond the scope of this book, we shall briefly illustrate the method.

Consider the electric dipole illustrated in Fig. 5.24. We assume that the current and charge of the dipole are given. Energy is radiated into space and at a distance the fields are as indicated in the figure so as to form a spherical wave. The radiating field components E_θ and H_ϕ are in phase with each other and their magnitudes vary inversely as the radius since the energy passing any spherical surface is constant and independent of the radius. E_θ and H_ϕ are related by the usual radiation coefficient so that $E_\theta = (\mu_0/\varepsilon_0)^{1/2} H_\phi$.

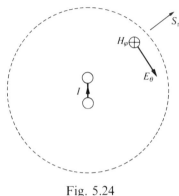

Fig. 5.24

Consider the variational principle of equation (3.106) and insert into it a harmonic time variation and adjoint quantities. This gives

$$-\langle \mathbf{H}^*, j\omega\, \delta\mathbf{B}\rangle + \langle j\omega\mathbf{D}^*, \delta\mathbf{E}\rangle - [\delta\mathbf{E} \times \mathbf{H}^*, \hat{\mathbf{n}}] - \langle \mathbf{J}^*, \delta\mathbf{E}\rangle = 0. \qquad (5.54)$$

Since we seek the radiation resistance, we can omit the j terms which are related to the stored energy. Let $\mathbf{J}^* = \bar{\mathbf{J}}^*$ be the impressed current at the dipole. Then

$$-\langle \bar{\mathbf{J}}^*, \delta\mathbf{E}\rangle - \left[\delta\mathbf{E}, \left(\frac{\varepsilon_0}{\mu_0}\right)^{1/2} \mathbf{E}^*\right] = 0. \qquad (5.55)$$

This can be integrated to give the variational statement

$$\delta\left[\langle \mathbf{J}^*, -\mathbf{E}\rangle - \frac{1}{2}\left\{\left(\frac{\varepsilon_0}{\mu_0}\right)^{1/2} E^2\right\}\right] = 0. \qquad (5.56)$$

Now we choose

$$E_\theta = \frac{k}{r}\exp\{j(\omega t - \beta r)\}\sin\theta$$

as a possible trial function where $\beta = \omega/c$ and k is an unknown coefficient. The surface integral is

$$\frac{1}{2}\int_0^{2\pi}\int_0^\pi \left(\frac{\varepsilon_0}{\mu_0}\right)^{1/2}\frac{k^2}{r^2}\sin^2\theta\, r\, d\theta r \sin\theta\, d\phi = \frac{4}{3}\pi\left(\frac{\varepsilon_0}{\mu_0}\right)^{1/2}k^2. \qquad (5.57)$$

The volume integral has to be taken close to the dipole, which is at the origin of co-ordinates.

The trial function in this region is

$$E = \frac{k}{r}\left(1 - j\beta r - \frac{\beta^2 r^2}{2}\right) \sin\theta$$

and $\sin\theta = 1$ in the direction of the current. Now the radiated power is finite and hence the electric field associated with this power must be finite as $r \to 0$. Thus we choose $E_\theta = -jk\beta$. If l is the length of the dipole and the current is constant along this length, we can write equation (5.56) as

$$\delta\left\{jk\beta Il - \frac{4}{3}\pi\left(\frac{\varepsilon_0}{\mu_0}\right)^{1/2} k^2\right\} = 0 \tag{5.58}$$

which gives

$$k = j\frac{3}{8\pi}\left(\frac{\varepsilon_0}{\mu_0}\right)^{1/2}\beta Il$$

and the radiated power is

$$\frac{3}{16\pi}\left(\frac{\varepsilon_0}{\mu_0}\right)^{1/2}\beta^2 I^2 l^2$$

By putting $\beta = 2\pi/\lambda$, where λ is the electromagnetic wavelength, we obtain the approximate radiation resistance as

$$R_+ = 1776(l/\lambda)^2. \tag{5.59}$$

The exact answer is $R = 790(l/\lambda)^2$, so the approximation is not very close. Let us seek to improve it by improving the trial function. We chose this from considerations of the distant field without paying much attention to the field near the dipole. This suggests that a better choice might be

$$E_\theta = k\left(\frac{1}{r^2} + \frac{\beta}{r}\right)\exp\left\{j(\omega t - \beta r)\right\} \sin\theta$$

where we have taken some account of the singularity at $r = 0$. We note that $\beta = 2\pi/\lambda$, so that $1/r^2$ and β/r are dimensionally the same. We now have for the field near the dipole

$$E_\theta = k\left(\frac{1}{r^2} + \frac{\beta}{r}\right)\left(1 - j\beta r - \frac{\beta^2 r^2}{2} \ldots\right).$$

The finite term is now

$$E_\theta = -j\beta^2 - \beta^2/2.$$

This phase difference draws attention to the need for a further improvement of the trial function. The term in $1/r^2$ was introduced to model the local field. Hence it must be dominantly the inductance field which lags the current by $\pi/2$, whereas the resistance field is in antiphase. Thus we should use

$$E_\theta = k\left(+\frac{j}{r^2} - \frac{\beta}{r} \right)\left(1 - j\beta r - \frac{\beta^2 r^2}{2} \cdots \right)$$

which gives a finite term

$$E_\theta = k\left(j\beta^2 - j\frac{\beta^2}{2} \right) = \tfrac{1}{2}j\beta^2 k.$$

This gives a new value for the radiation resistance as

$$R_+ = 888(l/\lambda)^2 \tag{5.60}$$

which is within 12.5 per cent of the exact value. Further improvement can be obtained by modelling the field close to the dipole more accurately. The method can in fact be used to find representations for singularities. The dual bound is more difficult to find, because it is more difficult to specify the voltage in a radiation problem. However, using our experience we can choose a trial function for H as

$$H_\phi = K\left(\frac{j}{r^2} - \frac{\beta}{r} \right) \exp\{j(\omega t - \beta r)\} \sin \theta.$$

Now we know the field of a current element to be

$$H_\phi = \frac{Il}{4\pi r^2} \sin \theta$$

as the frequency tends to be zero. Hence we can put $jk = Il/4\pi$. The radiated power is

$$-\frac{4}{3}\pi\left(\frac{\varepsilon_0}{\mu_0}\right)^{1/2} k^2\beta^2 = \frac{4}{3}\pi\left(\frac{\mu_0}{\varepsilon_0}\right)\frac{I^2 l^2}{(4\pi)^2}\frac{4\pi^2}{\lambda^2}$$

and this gives

$$R_- = 790(l/\lambda)^2. \tag{5.61}$$

In this case the trial function gives the actual field and thus the correct resistance.

We have worked this example in considerable detail to show the problems involved in applying the method. An alternative solution could have been found by selecting trial functions for the vector and scalar potentials. We can then use equations (3.122) and (3.126) and this may be helpful if the input admittance of an antenna is to be calculated.

The dipole is the basic building brick of antenna systems and the total energy of a system can be derived from a consideration of the mutual energy of dipoles.

5.12. Transmission line calculations

The electromagnetic equations of an ideal lossless transmission line have a particularly simple form which is called the TEM mode, because the electric and magnetic field vectors lie in a plane which is transverse, or perpendicular, to the direction of propagation.

Let the transverse plane be described by the rectangular co-ordinates x, y and let the wave solution be of the form $e^{j(\omega t - jz)}$. The field components will be E_x, E_y, H_x, and H_y. The curl equations are

$$\mathbf{V} \times \mathbf{E} = -j\omega\mu\mathbf{H} \qquad (5.62)$$

$$\mathbf{V} \times \mathbf{H} = j\omega\varepsilon\mathbf{E}. \qquad (5.63)$$

Hence

$$\frac{\partial E_y}{\partial x} - \frac{\partial E_x}{\partial y} = 0 \qquad (5.64)$$

and

$$\frac{\partial H_y}{\partial x} - \frac{\partial H_x}{\partial y} = 0. \qquad (5.65)$$

Thus E_y, E_x, H_y, and H_x can be expressed in terms of the gradient of a scalar potential function of x and y multiplied by the exponential propagation function of z and t.

$$\mathbf{E} = -\mathbf{V}\phi \, e^{j(\omega t - jz)} \qquad (5.66)$$

$$\mathbf{H} = -\mathbf{V}\psi \, e^{j(\omega t - jz)}. \qquad (5.67)$$

The potentials ϕ and ψ are related by Maxwell's equations. The problem of finding $\mathbf{E}$ and $\mathbf{H}$ is now greatly simplified, because it is very similar to the electrostatic and magnetostatic problem in which a scalar potential defines the solution.

This greatly enlarges the usefulness of the energy methods for static fields which we have discussed in this chapter in §§ 5.4, 5.5, 5.7, and 5.8. A full discussion is beyond the scope of this book, in which we are chiefly concerned with basic principles and very simple applications. The interested reader is referred to the excellent book by R. E. Collin: *Field Theory of Guided Waves*. In that book dual bounds are calculated for the capacitance and hence the characteristic impedance of a stripline (p. 164) and also for a capacitance discontinuity in a transmission line (p. 346). The TEM solution implies lossless propagation and becomes unreliable at frequencies above a few GHz. But within these restrictions it provides a powerful tool for the solution of transmission line problems.

Further reading

The book by R. E. Collin, *Field Theory of Guided Waves* (McGraw-Hill, New York, 1960), gives a number of bounded solutions for microwave striplines and wave-guides. It also has a very full account of the theory behind these solutions.

An extensive treatment of electromagnetic radiation by energy methods is given in the book *Field computation by moment methods* by R. F. Harrington (Macmillan, London, 1968). This book is especially valuable for its numerical examples.

6. Electromechanical Energy Conversion

6.1. Calculation of force and torque

In the previous chapters of this book we have looked at electromagnetic systems in terms of Lagrangian mechanics. We have become familiar with the application of the powerful concepts of energy and equilibrium to electromagnetic quantities and it has become obvious that the language of mechanics can easily incorporate electrical and magnetic concepts. This results in a remarkably fruitful interchange between electrical and mechanical notions. For instance it enables us to model mechanical systems by electromagnetic systems and *vice versa*. Two branches of engineering thus interact with each other to the advantage of both.

However, more remains to be said, because so far we have dealt with the same ideas in both types of system but regarded the systems as different. For instance we have defined isolated electromagnetic systems purely in terms of electromagnetic quantities. We have done this deliberately, although we were aware that electromagnetic systems are always subsystems of some larger entity. Earnshaw's theorem, which we discussed in § 4.4, provided a forceful reminder that electrostatic systems for instance are inherently unstable and require mechanical constraints to stabilize them. Of course these mechanical forces are themselves electrical in origin, but the electrical forces are atomic and molecular and are not part of the subsystem defined by forces of the inverse square law of electrostatics.

This interaction between electrical and mechanical systems becomes most clearly visible in the study of devices which transform electrical to mechanical energy or *vice versa*. Lagrangian mechanics takes this interaction in its stride because it operates with general co-ordinates which can be of any kind whatever as long as they fit into a system which also embodies suitable velocities and momenta.

Consider for example the system of Fig. 6.1 which shows a capacitor in which one plate can move under the control of a force and f is the force exerted by the plate. The mechanical co-ordinate is x. For the electrical subsystem we can choose either I or V as the applied force. The former choice gives the charge Q as the momentum and V as the velocity. The latter

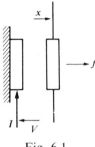

Fig. 6.1

gives Q as the co-ordinate and I as the velocity. The dual description is given in Table 6.1. The first formulation gives a kinetic energy

$$T = \int_0^V Q \, dV \tag{6.1}$$

and the dual gives a potential energy

$$U = \int_0^Q V \, dQ. \tag{6.2}$$

The force can be obtained by noting that T and U are functions of x but not of $\dot{x}$. Hence the use of the Euler–Lagrange equation (equation (2.17)) gives the force as

$$f_x = +\left(\frac{\partial T}{\partial x}\right)_V \tag{6.3}$$

and

$$f_x = -\left(\frac{\partial U}{\partial x}\right)_Q. \tag{6.4}$$

In each case the variation must be carried out with regard to the co-ordinate x while holding the other co-ordinate steady.

TABLE 6.1

q	$\dot{q}$	p	$-\dot{p}$
$\int V \, dt$	V	Q	$-I$
Q	I	$\int V \, dt$	$-V$

If the capacitance is independent of V or I we have $Q = VC$ and

$$T = U = \tfrac{1}{2}QV. \tag{6.5}$$

Hence the magnitude of f can be obtained without paying attention to the type of energy. The sign of the force is, however, different in the two cases because as always the kinetic energy is increased and the potential energy is decreased. In the kinetic energy case there is an energy interchange with the source and in the potential energy case the source is unaffected by the variational process. It is essential to specify the type of energy and this is done by specifying the co-ordinate which is held constant. The statement $f = \partial W/\partial x$, where $W = \tfrac{1}{2}QV$ is the 'stored energy', has no meaning.

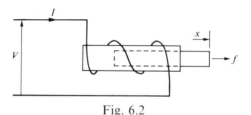

Fig. 6.2

Moreover, if the relationship between V and Q is non-linear even the magnitude of the two types of energy is different in accordance with the choice of electrical co-ordinate. We shall discuss this with reference to the iron-cored inductance shown in Fig. 6.2.

The electrical variables can be arranged as shown in Table 6.2. Φ is the flux linkage. In terms of Φ and I we have

$$T = \int_0^I \Phi \, dI \tag{6.6}$$

$$U = \int_0^\Phi I \, d\Phi. \tag{6.7}$$

TABLE 6.2

q	$\dot{q}$	p	$-\dot{p}$
Φ	V	Q	$-I$
Q	I	Φ	$-V$

The force f_x is given by

$$f_x = + \left(\frac{\partial T}{\partial x} \right)_I \tag{6.8}$$

and

$$f_x = - \left(\frac{\partial U}{\partial x} \right)_\Phi . \tag{6.9}$$

If the relationships between Φ and I is as indicated in Fig. 6.3, T and U are very different in magnitude and the correct energy must be chosen to give the correct value of the force. It is of great interest to notice that for a

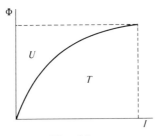

Fig. 6.3

linear device the stored energy U or T is equal to half the input of energy supplied at constant force. This means that in an energy conversion process under these conditions half the energy is stored and half is available as output. In the non-linear device, however, there is no such equipartition of energy. In the iron-cored inductance characterized by Fig. 6.3 U can be made very much smaller than T. If we consider the energy equation of the system

$$\int_0^t VI \, dt = \int_0^t I \frac{d\Phi}{dt} \, dt = \int_0^\Phi I \, d\Phi = \int_0^U dU + \int_0^w dw \tag{6.10}$$

where w is the work done on the environment. If the input occurs at constant current I, the process is defined and we can integrate

$$I\Phi = U + w. \tag{6.11}$$

Reference to Fig. 6.3 shows that U can be made small if there is saturation. Hence as $U \rightarrow 0$ the work done becomes almost equal to the electrical input. This means that the force or torque can be almost doubled as between a linear and a saturated device. It should be noted that the process has to be carried out at constant current. At constant flux the energy relationship is

$$w = -\Delta U. \tag{6.12}$$

Reference to Fig. 6.3 shows that the mechanical output becomes small because ΔU is very small at constant flux. It needs hardly to be said that saturation cannot provide an increase of output in a cyclical process. An interesting discussion of the effect of saturation in increasing force is given by Byrne.†

So far we have dealt with an ideal energy conversion process which has no dissipation. Suppose the coil in Fig. 6.2 has electrical resistance R; we then have

$$V - IR = \frac{d\Phi}{dt} \tag{6.13}$$

and V can no longer be treated as one of the Lagrangian velocities or forces. This means that V cannot be derived from the energy by differentiation in the Euler–Lagrange equations. However, we can invent an internal voltage $E = V - IR$ which can be derived in this manner. The procedure to be adopted, therefore, is to calculate the energies for a conservative system from which the dissipative forces have been removed. Then the conservative forces can be calculated by differentiation and lastly the non-conservative forces can be introduced. This is the same procedure as is required for non-holonomic constraints which were discussed in § 2.3. The difficulty in the treatment of non-holonomic constraints arises because it is not possible to integrate their virtual work. Similarly the dissipative work terms cannot be integrated. A dissipative system cannot be treated as an isolated system in equilibrium and this is the fundamental basis of Lagrangian mechanics. We overcame this difficulty in our consideration of the diffusion equation in § 3.3.4 by introducing an adjoint system which generated energy at the same rate at which it was being dissipated. This could be done here as well, but in electromechanical energy conversion it is generally convenient to deal with the problem in two stages. First the system parameters such as electrical resistance, inductance, capacitance, mechanical friction, inertia and spring stiffness are calculated and then the system is treated as an assemblage of ideal elements. In Chapter 5 we gave an account of the calculation of the electrical parameters. In this chapter we assume that the parameters have been calculated and we are able to isolate the dissipative elements. The system is therefore a conservative system with non-conservative elements on its input and output terminals. These considerations underlie the distinction between potential difference and electromotive force in treating electrical machines. The former includes the dissipation in resistances and the latter is

† J. V. Byrne, Tangential forces in overlapped pole geometries incorporating ideally saturable material, *IEEE Trans. Magn.* **8** (1), 2–9 (1972).

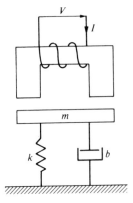

Fig. 6.4

a conservative force. The reason for this distinction is seldom given explicitly nor is it often stated that the removal of the dissipative processes to the machine terminals limits the usefulness of this representation for the understanding of processes inside the machine. Investigations involving such phenomena as parasitic currents, for instance, require a more careful subdivision of the system and may derive great benefit from a variational method involving an adjoint system.

We conclude this general discussion by examining the particular system illustrated in Fig. 6.4. The system consists of an electromagnet which is energized through a coil having resistance R. The magnet acts on a moving iron element of mass m supported by a spring of constant k and damped by a dashpot of coefficient b. It is required to find the equation of motion of the system.

We consider first the conservative part of the system and choose to represent the magnetic energy as kinetic energy. This gives the choice of co-ordinates and momenta given in Table 6.3. The effect of gravity can be included in the setting of the spring, i.e. in the datum level of x.

The Lagrangian is given by

$$L = \tfrac{1}{2}m\dot{x}^2 + \int \Phi \, dI - \tfrac{1}{2}kx^2 + \int E \, dQ. \tag{6.14}$$

TABLE 6.3

q	$\dot{q}$	p	$-\dot{p}$
Q	I	Φ	$-E$
x	$\dot{x}$	$m\dot{x}$	$-kx$

The Euler–Lagrange equations are

$$m\ddot{x} - \frac{\partial}{\partial x}\left(\int \Phi \, dI\right)_I + kx = 0 \tag{6.15}$$

$$\frac{d\Phi}{dt} - E = 0. \tag{6.16}$$

We now include the dissipative elements

$$m\ddot{x} + kx + b\dot{x} - \frac{\partial}{\partial x}\left(\int \Phi \, dI\right)_I = 0 \tag{6.17}$$

$$\frac{d\Phi}{dt} - V + IR = 0. \tag{6.18}$$

The dual choice of the electromagnetic variables gives

$$L = \tfrac{1}{2}m\dot{x}^2 - \int Q \, dE - \tfrac{1}{2}kx^2 - \int I \cdot d\Phi \tag{6.19}$$

which gives for the equilibrium equations

$$m\ddot{x} + kx + \frac{\partial}{\partial x}\left(\int I \, d\Phi\right)_\Phi = 0 \tag{6.20}$$

and

$$-I + I = 0. \tag{6.21}$$

When the non-conservative elements are included we obtain

$$m\ddot{x} + b\dot{x} + kx + \frac{\partial}{\partial x}\left(\int I \, d\Phi\right)_\Phi = 0 \tag{6.22}$$

and

$$I_{\text{total}} = I + gV \tag{6.23}$$

where g is the dissipative conductance.

It is interesting to note that the dual transforms the loop to the node equation and that the dissipative element must therefore be inserted in parallel instead of in series.

6.2. The use of rotating axes in the analysis of electrical machines

The examples we have discussed in the previous section had a single coil, but the method is equally applicable to systems containing any number of coils. Suppose we have two coils and suppose the magnetic system is linear. We can write the flux linkages in terms of inductance coefficients and currents as

$$\begin{bmatrix} \Phi_1 \\ \Phi_2 \end{bmatrix} = \begin{bmatrix} L_{11} & L_{12} \\ L_{12} & L_{22} \end{bmatrix} \begin{bmatrix} I_1 \\ I_2 \end{bmatrix}. \tag{6.24}$$

To obtain the kinetic energy we have to consider the integral $\int \Phi \, dI$ for the system. Suppose we first take both currents as zero at our datum of zero energy and then integrate I_1 to its full value while keeping I_2 at zero and proceed with the integration by keeping I fixed and raising I_2 from zero to its full value; we then obtain

$$T = \tfrac{1}{2}L_{11}I_1{}^2 + L_{12}I_1I_2 + \tfrac{1}{2}L_{22}I_2{}^2$$

$$= \tfrac{1}{2}[I_1 I_2] \begin{bmatrix} L_{11} & L_{12} \\ L_{12} & L_{22} \end{bmatrix} \begin{bmatrix} I_1 \\ I_2 \end{bmatrix}. \tag{6.25}$$

Similarly the potential energy $\int I \, d\Phi$ takes the form

$$U = \tfrac{1}{2}[\Phi_1 \Phi_2] \begin{bmatrix} T_{11} & T_{12} \\ T_{12} & T_{22} \end{bmatrix} \begin{bmatrix} \Phi_1 \\ \Phi_2 \end{bmatrix} \tag{6.26}$$

where the T matrix is the inverse of the L matrix.

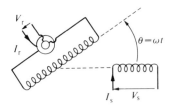

Fig. 6.5

Consider now the simple electrical machine illustrated by Fig. 6.5. This machine has a stationary coil designated by the letter s and a rotating coil designated by r. The device converts electrical to mechanical energy by means of a variable mutual inductance L_{sr}. The rotating coil is supplied through slip rings. We consider the conversion of electrical to mechanical energy and omit the dissipative forces, knowing that they can always be added to the differential equations. Let the polar moment of inertia of the rotating parts be J. Let the co-ordinates and momenta be as given in Table 6.4. The Lagrangian energy is given by

$$L = \tfrac{1}{2}J\omega^2 + \tfrac{1}{2}[I_s I_r] \begin{bmatrix} L_{ss} & L_{sr} \\ L_{sr} & L_{rr} \end{bmatrix} \begin{bmatrix} I_s \\ I_r \end{bmatrix} - \int T \, d\theta + \int V_s \, dQ_s + \int V_r \, dQ_r. \tag{6.27}$$

TABLE 6.4

q	$\dot{q}$	p	$-\dot{p}$
$\begin{bmatrix} Q_s \\ Q_r \end{bmatrix}$	$\begin{bmatrix} I_s \\ I_r \end{bmatrix}$	$\begin{bmatrix} \Phi_s \\ \Phi_r \end{bmatrix}$	$\begin{bmatrix} V_s \\ V_r \end{bmatrix}$
θ	ω	$J\omega$	$-T$

The differential equations are therefore given by

$$L_{ss}\frac{dI_s}{dt} + L_{sr}\frac{dI_r}{dt} + I_r\frac{dL_{sr}}{dt} - V_s = 0 \tag{6.28}$$

$$L_{rr}\frac{dI_r}{dt} + L_{sr}\frac{dI_s}{dt} + I_s\frac{dL_{sr}}{dt} - V_r = 0 \tag{6.29}$$

$$J\frac{d\omega}{dt} - I_s I_r\frac{dL_{sr}}{d\theta} + T = 0. \tag{6.30}$$

If $L_{sr} = M \cos\theta$, the torque equation becomes

$$J\frac{d\omega}{dt} + I_s I_r M \sin\theta + T = 0. \tag{6.31}$$

If the speed is constant

$$T = -I_1 I_2 M \sin\theta \tag{6.32}$$

where T is the generated torque.

This result can be generalized so that, for any number of windings, using equation (6.27) we have

$$T = \tfrac{1}{2}[I]_t \left[\frac{\partial L}{\partial\theta} \right] [I]. \tag{6.33}$$

It is convenient to treat multi-pole machines in terms of two-pole machines and to transform the effect of polyphase windings to equivalent two-phase windings. This ensures that the electrical and mechanical angles are the same, and it enables the fluxes to be expressed in terms of components along and perpendicular to this angle.

Let us consider a three-phase to two-phase transformation as illustrated in Fig. 6.6. Let the systems be balanced and let them be operating at a steady angular frequency ω. We have

$$\begin{aligned} I_a &= I_3 \cos\omega t \\ I_b &= I_3 \cos(\omega t - 2\pi/3) \\ I_c &= I_3 \cos(\omega t - 4\pi/3) \end{aligned} \tag{6.34}$$

and

$$I_\alpha = I_2 \cos \omega t$$
$$I_\beta = I_2 \cos (\omega t - \pi/2) = I_2 \sin \omega t \qquad (6.35)$$

where I_3 is the amplitude of the three-phase currents and I_2 the amplitude of the two-phase currents. I_3 and I_2 are related by the fact that they are to produce the same power. This means that the product of current and

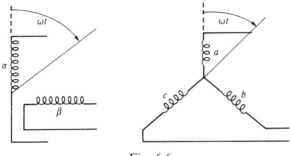

Fig. 6.6

magnetomotive force must remain the same in the transformation. On the assumption of the same number of turns we therefore have

$$I_2 = \sqrt{\tfrac{3}{2}} I_3 \qquad (6.36)$$

and

$$\begin{bmatrix} I_\alpha \\ I_\beta \end{bmatrix} = \sqrt{\tfrac{2}{3}} \begin{bmatrix} 1 & -\tfrac{1}{2} & -\tfrac{1}{2} \\ 0 & \sqrt{3}/2 & \sqrt{3}/2 \end{bmatrix} \begin{bmatrix} I_a \\ I_b \\ I_c \end{bmatrix}. \qquad (6.37)$$

The transformation to two-phase windings is useful in its application to rotating machines having a uniform magnetic structure such as induction motors. Synchronous and direct current machines, however, are designed with a structure which on either the stator or rotor exhibits differences between the direct axis which lies along a pole centre-line and the quadrature axis which lies along the line between poles. The flux therefore depends on this difference. A further transformation to axes fixed to the direct and quadrature system is highly convenient. If the direct and quadrature axes are fixed with respect to the stator, this means a transformation from rotating to fixed axes. We propose to investigate this transformation because it is 'non-holonomic' as was discussed in § 2.3. Such non-holonomic constraints need to be treated carefully in a Lagrangian scheme of calculation.

Consider a transformation from a rotating two-phase to a stationary two-axis system. Let the stationary system be denoted by d for direct axis and q for quadrature axis. The systems are illustrated in Fig. 6.7. The transformation is given by

$$\begin{bmatrix} I_d \\ I_q \end{bmatrix} = \begin{bmatrix} \cos\theta & \sin\theta \\ \sin\theta & -\cos\theta \end{bmatrix} \begin{bmatrix} I_\alpha \\ I_\beta \end{bmatrix}, \tag{6.38}$$

This transformation differs from that of equation (6.37), because $\theta = \omega t$ so that the transformation depends on time. The d and q coils are fictitious because the actual winding rotates and it is only the flux system which is stationary. The transformation could be brought about physically by attaching a commutator to the rotating windings. The d and q currents would then be the currents at the brushes on the commutator and they would not be the currents in the windings.

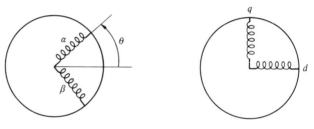

Fig. 6.7

Mathematically the transformation applies a constraint to the system which is expressed in a relation between the currents but not between the charges. In terms of the Lagrangian formulation there is a constraint on the velocities but not on the co-ordinates. Now although it is possible to obtain the velocities from the co-ordinates the reverse process is not always possible. The/commutator transformation cannot be integrated and is said to be non-holonomic. It is therefore not possible to find a Lagrangian energy in terms of the d and q currents.

However, the difficulty can be overcome by introducing the transformation in its differential form into the Euler–Lagrange equations. As an example consider a generator with rotating α, β windings and a stationary winding denoted by D as shown in Fig. 6.8. Let us consider the machine to be of the salient-pole type so that the inductances depend on the position of the coils relative to the direct and quadrature axes. Let us further assume that the inductances can be expressed in terms of a Fourier series in θ and that terms beyond the second harmonic can be neglected.

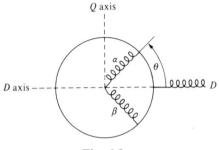

Fig. 6.8

Thus we write

$$L_{\alpha\alpha} = L_0 + L_1 \cos 2\theta. \qquad (6.39)$$

We intend to work in terms of equivalent d, q windings on the rotor and note that

$$L_d = L_0 + L_1 \qquad (6.40)$$

$$L_q = L_0 - L_1. \qquad (6.41)$$

Hence

$$L_0 = (L_d + L_q)/2 \qquad (6.42)$$

$$L_1 = (L_d - L_q)/2. \qquad (6.43)$$

The mutual inductance between the α and β windings caused by the saliency of the magnetic structure is given by

$$L_{\alpha\beta} = +L_1 \sin 2\theta. \qquad (6.44)$$

The mutual inductances between the α and D windings and the β and D windings are

$$L_{\alpha D} = M \cos \theta \qquad (6.45)$$

$$L_{\beta D} = M \sin \theta. \qquad (6.46)$$

Referring to equation (6.28) we can write

$$p\{\tfrac{1}{2}(L_d + L_q) + \tfrac{1}{2}(L_d - L_q) \cos 2\theta\}I_\alpha$$
$$+ p\{\tfrac{1}{2}(L_d - L_q) \sin 2\theta\}I_\beta + pM \cos \theta I_D = V_\alpha \qquad (6.47)$$

where $p \equiv d/dt$.

This can be simplified to

$$V_\alpha = p(L_d \cos^2 \theta + L_q \sin^2 \theta)I_\alpha$$
$$+ p(L_d - L_q) \sin \theta \cos \theta I_\beta + pM \cos \theta I_D. \qquad (6.48)$$

The trouble with this equation is that it contains θ which varies with time. A better form results if we introduce the non-holonomic transformation

$$\begin{bmatrix} I_\alpha \\ I_\beta \end{bmatrix} = \begin{bmatrix} \cos\theta & \sin\theta \\ \sin\theta & -\cos\theta \end{bmatrix} \begin{bmatrix} I_d \\ I_q \end{bmatrix} \tag{6.49}$$

$$\begin{aligned} V_\alpha &= p(L_d \cos^2\theta + L_q \sin^2\theta)(I_d \cos\theta + I_q \sin\theta) \\ &\quad + p(L_d - L_q)\sin\theta\cos\theta(I_d \sin\theta - I_q \cos\theta) \\ &\quad + pM\cos\theta I_d \\ &= p(L_d I_d \cos\theta) + p(L_q I_q \sin\theta) + pMI_D \cos\theta. \end{aligned} \tag{6.50}$$

If we now introduce a similar transformation for the voltages

$$\begin{bmatrix} V_\alpha \\ V_\beta \end{bmatrix} = \begin{bmatrix} \cos\theta & \sin\theta \\ \sin\theta & -\cos\theta \end{bmatrix} \begin{bmatrix} V_d \\ V_q \end{bmatrix} \tag{6.51}$$

we obtain

$$\begin{aligned} V_d \cos\theta + V_q \sin\theta &= L_d \cos\theta pI_d - L_d I_d \sin\theta\dot\theta \\ &\quad + L_q \sin\theta pI_q + L_q I_q \cos\theta\dot\theta \\ &\quad + M\cos\theta pI_D - MI_D \sin\theta\dot\theta. \end{aligned} \tag{6.52}$$

Equating terms in $\cos\theta$ and $\sin\theta$ we have

$$\begin{bmatrix} V_d \\ V_q \end{bmatrix} = \begin{bmatrix} L_d p & L_q \dot\theta & Mp \\ -L_d \dot\theta & L_q p & -M\dot\theta \end{bmatrix} \begin{bmatrix} I_d \\ I_q \\ I_D \end{bmatrix}. \tag{6.53}$$

Similarly we can obtain the equation for V_D from consideration of equation (6.27). This gives

$$V_D = L_D pI_D + MpI_d. \tag{6.54}$$

Hence the complete set of equations is

$$\begin{bmatrix} V_d \\ V_q \\ V_D \end{bmatrix} = \begin{bmatrix} L_d p & L_q \dot\theta & Mp \\ -L_d \dot\theta & L_q p & -M\dot\theta \\ Mp & - & L_D p \end{bmatrix} \begin{bmatrix} I_d \\ I_q \\ I_D \end{bmatrix}. \tag{6.55}$$

Reference to equations (6.27) and (6.30) shows that the torque depends only on the rotational terms. We see from equation (6.55) that the transformation to d, q coils has resulted in an expression for the torque of the form

$$T = [I]_t [G][I] \tag{6.56}$$

where G is the part of the impedance matrix which has terms depending on the rotation. In our example

$$T = [I_d I_q I_D] \begin{bmatrix} - & L_q & - \\ -L_d & - & -M \\ - & - & - \end{bmatrix} \begin{bmatrix} I_d \\ I_q \\ I_D \end{bmatrix}$$

$$= -MI_q I_D - (L_d - L_q)I_d I_q \tag{6.57}$$

The first term gives the interaction of the q and D currents and the second the effect of saliency of the magnetic structure. The sign depends on the choice of the q winding with respect to the direction of rotation.

This example illustrates the great benefit derived from the transformation to d, q coils in eliminating the troublesome angle of rotation from the voltage, current and torque expressions. We have already pointed out that the transformation is non-holonomic and must be applied after the Euler–Lagrange differential equations have been derived from the Lagrangian energy. It is instructive to note what happens if the order is reversed and if the transformation is applied to the energy first. Let us assume, erroneously, that the currents I_d and I_q can be associated with charges Q_d, Q_q which obey the transformation

$$\begin{bmatrix} Q_\alpha \\ Q_\beta \end{bmatrix} = \begin{bmatrix} \cos\theta & \sin\theta \\ \sin\theta & -\cos\theta \end{bmatrix} \begin{bmatrix} Q_d \\ Q_q \end{bmatrix}. \tag{6.58}$$

We now attempt to write the Lagrangian of equation (6.27) in terms of the transformed parameters:

$$L = \tfrac{1}{2}J\omega^2 + \tfrac{1}{2}L_d I_d{}^2 + \tfrac{1}{2}L_q I_q{}^2 + \tfrac{1}{2}L_D I_D{}^2 + MI_d I_D$$

$$+ V_d Q_d + V_q Q_q + V_D Q_D - \int T\,d\theta. \tag{6.59}$$

We see immediately that only the torque integral and the rotational energy depend on the angle of rotation. The electrical terms are all independent of θ. Hence the voltage equations have no rotational terms and the generated torque is independent of the electrical energy. Clearly equation (6.59) is incorrect because the energy is not invariant under a non-holonomic transformation.

This difficulty has led some writers to speak of a failure of the Lagrangian method and to contrast this method with methods which derive the differential equations heuristically by a superposition of various terms. As we have noticed, however, the Lagrangian method can be applied with confidence once it is noted that equation (6.38) describes a transformation which cannot be integrated to give equation (6.58). The Langrangian method draws attention to the fact that I_d and I_q are equivalent currents. Thus the mathematical and the physical descriptions are in close accord in the Lagrangian scheme, which is therefore particularly well adapted for the study of all kinds of electrical machines.

6.3. Parametric energy processes

At various stages in this book we have touched on the application of variational methods to non-linear systems. Thus in the discussion of the variational principles applied to electrical networks in § 3.2 we saw that Tellegen's theorem does not require linear relationships between the voltages and currents of the elements of the system. Similarly in distributed field problems a great advantage of variational methods is that non-linear problems present no special difficulty. It is even possible to obtain doubly bounded solutions subject to the provision that negative slopes in the transfer characteristics are excluded. This was discussed in § 3.3.1. Clearly that discussion, which was in the context of electrostatics, applies equally well to magnetostatic problems.

However, Tellegen's theorem deals with instantaneous currents and voltages and our discussion of non-linear field problems also only dealt with static fields. We must now consider time-varying non-linear energy processes. This is likely to be a very difficult matter because the non-linearity will give rise to many frequencies. On the other hand the problem is of great practical interest because non-linear devices can be used as frequency convertors. Power can be fed into the device at one frequency and extracted at another frequency. Such processes are generally called parametric energy processes. We seek to analyse them in the context of variational energy principles.

We shall be guided by our experience of the behaviour of time-varying systems in § 3.3.4. There we saw that in order to establish a time-invariant energy we had to deal with an adjoint system which was the complex conjugate of the system under discussion. The interaction of the two systems was then independent of time. In our present discussion we face the additional difficulty of having to deal with a large number of frequencies. It is likely, therefore, that we shall have to use a Fourier series for the frequencies and an energy expression which is made up of a number of frequency-dependent terms.

Let us consider this in the context of a non-linear capacitance as illustrated in Fig. 6.9 where V is the potential difference and Q the charge on the plates of the capacitor. In the Lagrangian description we choose Q as the generalized co-ordinate and V as the applied force.

Let V and Q be given in complex form by

$$V = V_0 + \sum_n V_n \exp \{j(\omega_n t + \phi_n)\} \tag{6.60}$$

$$Q = Q_0 + \sum_n Q_n \exp \{j(\omega_n t + \lambda_n)\} \tag{6.61}$$

Hence the average energy can be written

$$|VQ| = V_0 Q_0 + \tfrac{1}{2} \sum_n V_n Q_n{}^* \tag{6.62}$$

where we have incorporated the phase angles ϕ_n and λ_n in what are now the complex expressions for the voltage and charge components. $Q_n{}^*$ is the complex conjugate of Q_n. Equation (6.62) is a complex quantity and

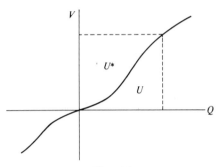

Fig. 6.9

contains real and imaginary terms. Since we are used to dealing with real and reactive power rather than energy we can substitute current for charge by the expressions

$$I_n{}^* = -j\omega_n Q_n{}^*. \tag{6.63}$$

Hence

$$|VQ| = V_0 Q_0 + \frac{1}{2} j \sum_n \frac{V_n I_n{}^*}{\omega_n}. \tag{6.64}$$

If we define the real power components by

$$P_n = \mathrm{Re}\left(\frac{V_n I_n{}^*}{2}\right) \tag{6.65}$$

and the reactive power components by

$$R_n = -\mathrm{Im}\left(\frac{V_n I_n{}^*}{2}\right) \tag{6.66}$$

where the sign of the reactive power is chosen as negative for a capacitor, we obtain

$$|VQ| = V_0 Q_0 + \sum_n \frac{(jP_n + R_n)}{\omega_n} \tag{6.67}$$

and for a loss-free capacitor we also know that

$$\sum_n P_n = 0. \tag{6.68}$$

Other useful results can be obtained by applying the variational method. Consider the virtual work

$$\text{Re } V \, \delta Q^* = 0 \tag{6.69}$$

where the arbitrary variation δQ^* is to be carried out in terms of arbitrary changes in the independent phase angles:

$$V \, \delta Q^* = \sum_n V_n \, \delta Q_n^* = \sum_n j \frac{V_n}{\omega_n} \delta I_n^*. \tag{6.70}$$

Now the variation affects the phase angles of the I_n components but it leaves the explicit time dependence unchanged because I_n^* has to be adjoint in time to V_n. The virtual work statement of equation (6.69) is applied to the average and not to the instantaneous energy. The variation of the phase angles can be written as

$$\delta \exp (j\theta_n) = j \frac{\partial \theta_n}{\partial \theta_k} \delta \theta_k \tag{6.71}$$

where k is an index which describes the independent phase angles on which the θ_n depend. However, we seek to express the variation in terms of frequencies. Suppose

$$\theta_n = f(\omega_k). \tag{6.72}$$

Then we can write

$$\delta \exp (j\theta_n) = j \frac{\partial \theta_n}{\partial \omega_k} \delta \omega_k \tag{6.73}$$

where the dependence of the θ_n on the ω_k involves the shape of the V, Q characteristic.

Hence

$$V \, \delta Q^* = \sum_n \frac{V_n I_n^*}{\omega_n} \delta \omega_k \frac{\partial \theta_n}{\partial \omega_k}. \tag{6.74}$$

However, this relationship is independent of the shape of the V, Q curve. Hence finally

$$\text{Re} \sum_n \frac{V_n I_n^*}{\omega_n} \delta \omega_k = \sum \frac{P_n}{\omega_n} \delta \omega_k = 0. \tag{6.75}$$

In a system in which the free ω_k are ω_1 and ω_0 and the dependent frequencies are given by $m\omega_1 + n\omega_0$, we can derive from equation (6.75):

$$\sum_{m=1}^{\infty} \sum_{n=-\infty}^{\infty} \frac{mP_{mn}}{m\omega_1 + n\omega_0} = 0 \qquad (6.76)$$

$$\sum_{m=-\infty}^{\infty} \sum_{n=1}^{\infty} \frac{nP_{mn}}{m\omega_1 + n\omega_0} = 0. \qquad (6.77)$$

These equations are known after their originators as the Manley–Rowe equations.

Our proof has been based on a variational principle applied to the energy stored in the non-linear capacitor. This principle asserts that this energy has an equilibrium value which can be explored by a small displacement.

In our previous discussion of the energy stored in electrostatic fields we were able to find dual principles and upper and lower bounds to approximate solutions. These depended on the second variation of the energy and on the convexity or concavity of the energy functionals. It is of great interest to find out whether such an analysis can be extended to the time-varying non-linear case which we are discussing in this section.

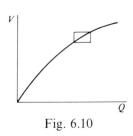

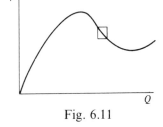

Fig. 6.10   Fig. 6.11

Consider the two transfer characteristics shown in Figs. 6.10 and 6.11. Consider the small second-order energy $\delta V\,\delta Q$. In Fig. 6.10 this energy is essentially positive, but in Fig. 6.11 there are regions where a δQ is associated with a negative δV. We can therefore make a statement about the second-order variation associated with characteristics of the type of Fig. 6.10 by requiring that

$$\delta V\,\delta Q \geqslant 0. \qquad (6.78)$$

In the time-varying case we can make a statement about the average energy variation

$$\mathrm{Re}\,\frac{1}{2}\sum_n \delta V_n\,\delta Q_n{}^* = \mathrm{Re}\,\frac{1}{2}\sum \delta V_{nj} \frac{\delta I_n{}^*}{\omega_n}$$

$$= I_m\,\frac{1}{2}\sum \frac{\partial V_n\,\delta I_n{}^*}{\omega_n}. \qquad (6.79)$$

In accordance with the conventional sign of the reactive power in a capacitor this energy must be negative or equal to zero.

The variations δV_n, $\delta I_n{}^*$ are carried out in the manner we have discussed for the first variation above, and we obtain

$$\sum_n \frac{R_n}{\omega_n} (\delta \omega_k)^2 \leqslant 0. \tag{6.80}$$

The discussion for non-linear inductance is exactly the same as for non-linear capacitance except that the sign of the inequality in equation (6.80) is changed. Dual principles can be obtained as in the static case by using approximate spatial distributions of one or other of the parameters in the transfer characteristics.

We now discuss the non-linear resistance. The variational principle in such an application is based on the equilibrium condition for the power loss which is often called the content. The transfer characteristic has the variables voltage and current, so that the current takes the place of the charge in the previous discussion of the capacitor. Also the role of the active and reactive power is interchanged. We can therefore write immediately that

$$\sum_n R_n \delta \omega_k = 0 \tag{6.81}$$

where we note that the R_n are not divided by the frequencies ω_n since the virtual work becomes a virtual power principle. Similarly we can write

$$\sum_n P_n (\delta \omega_k)^2 \geqslant 0. \tag{6.82}$$

In spite of the eloquent advocacy of Penfield in his book *Frequency–power formulas*, details of which are given at the end of this chapter, the usefulness of these principles does not seem to have been appreciated very generally. Our brief treatment certainly can serve only as an appetizer. The reason for including our discussion in this book is that the Manley–Rowe formulae and associated formulae fall naturally into Lagrangian mechanics and provide an exceedingly powerful extension for the analysis of non-linear systems by means of Fourier series. It is to be hoped that the application of these equations will be extended. Moreover the method of proof suggests that other orthogonal expansions besides the Fourier expansion could also be used in variational principles where appropriate.

We conclude this section with some simple applications.

Suppose we have a simple reactive system in which the power flow is associated with three frequencies ω_0, ω_1 and a sideband $\omega_n = \omega_1 + n\omega_0$.

Equations (6.76) and (6.77) give

$$\frac{P_{10}}{\omega_1} + \frac{P_{1n}}{\omega_n} = 0 \tag{6.83}$$

$$\frac{P_{01}}{\omega_0} + \frac{nP_{1n}}{\omega_n} = 0. \tag{6.84}$$

From equation (6.83)

$$\frac{P_{1n}}{P_{10}} = -\frac{\omega_n}{\omega_1}. \tag{6.85}$$

Hence a power input P_{10} results in a power output P_{1n} if ω_n is positive. Such a device acts as an upper-sideband converter. If on the other hand ω_n is negative equation (6.83) shows that P_{10} and P_{1n} have the same sign. Hence the power put in at the frequency ω_0, which is P_{01}, can be taken out at ω_1 and at ω_n. The interested reader should consult the book by Howson and Smith, details of which are given at the end of this chapter.

Rotating electrical machines can be treated as non-linear inductances subject to the condition that dissipative elements such as the winding resistances and other losses must be removed to the terminals of the inductance. Moreover care must be taken in the application of the commutator transformation which we have discussed in § 6.2. An interesting example discussed by Penfield is the balanced induction machine. Let the subscripts s and r denote the stator and rotor windings and let the shaft speed be given by a constant ω_m. If ω_s and ω_m are the independent frequencies equations (6.76) and (6.77) give

$$\frac{P_m}{\omega_m} - \frac{P_r}{\omega_r} = 0 \tag{6.86}$$

and

$$\frac{P_s}{\omega_s} + \frac{P_r}{\omega_r} = 0. \tag{6.87}$$

The stator frequency ω_s is positive but ω_r and ω_m can be either positive or negative. We have defined the rotor frequency as

$$\omega_r = \omega_s - \omega_m. \tag{6.88}$$

We assume that all the powers are defined in terms of input. Then

$$P_r = -\frac{\omega_r}{\omega_s} P_s \tag{6.89}$$

and

$$P_m = -\frac{\omega_m}{\omega_s} P_s \tag{6.90}$$

which gives the well-known result that the input power divides itself into a mechanical output and an input to the rotor in the ratio of the rotational frequency to the rotor frequency. We notice that the rotor power is dissipated in the rotor resistance, but from the point of view of the stator inductance the dissipation is outside the system. The generating mode can be studied by putting ω_r negative and ω_m positive. This results in a negative P_s being associated with a positive input to the shaft of P_m. If P_m is positive and P_s is also positive the machine acts as a brake. This requires that ω_m is opposite in sign to ω_s.

A parametric process important in electrical machines has been investigated from first principles by Jordan *et al.*†. It concerns the saturation harmonics and the power associated with these harmonic fields. Consider an induction machine driven at synchronous speed which has a third harmonic in its field-form curve. Let P_s be the power supplied electrically to the stator at the frequency ω and let P_m be the power supplied mechanically to the shaft. Let the iron losses be P_1 at the frequency ω and P_3 at the frequency 3ω. These losses are assumed to take place in two separate stator windings, one having the number of poles associated with the fundamental frequency and the other having three times that number. The mechanical losses are ignored or at least excluded from the system under consideration.

The Manley–Rowe equations for this system can be written as

$$\frac{P_s}{\omega} - \frac{P_1}{\omega} - \frac{P_3}{3\omega} = 0 \qquad (6.91)$$

$$\frac{P_m}{\omega} - \frac{2}{3\omega} P_m = 0. \qquad (6.92)$$

Thus two-thirds of the saturation harmonic power are supplied from the mechanical torque and one-third from the stator input. The example shows the power of the method in dealing with complicated non-linear phenomena in a very straightforward manner.

Further reading

A detailed discussion of the calculation of torque in electrical machines is given in the book *Matrix analysis of electrical machinery*, N. N. Hancock, Pergamon Press, Oxford (1974).

The Manley–Rowe equations are studied fully in the book *Frequency–power formulas*, P. Penfield, MIT/Wiley, New York (1960).

Parametric amplifiers are described in the book *Parametric amplifiers*, D. P. Howson and R. B. Smith, McGraw-Hill, New York (1970).

† H. Jordan, V. Klima, W. Raube, and G. Röder, Über parametrisch erzeugte Oberfelder, *E.T.Z.-A.*, **92**, 451–455 (1971).

7. Postscript: A Comparison of Methods for the Calculation of Electromagnetic Fields

7.1. The importance of experience

The range of electromagnetic problems is very large and beyond the knowledge of all but a handful of exceptional people. Although it is true that all electromagnetic phenomena are governed by Maxwell's equations, the converse is not true inasmuch as knowledge of these equations does not confer understanding of all the phenomena. Until the equations are placed in the context of some physical process they remain abstract and symbolic. Engineers working on arc phenomena in switchgear, on magnetic fields in turbogenerators, on microwave links, or on computer hardware need much else besides a knowledge of the field equations. There is a necessary tension between the universality of the theoretical framework and the speciality of the application.

This tension exists also in the methods of field calculation. The same equations can be solved by a wide variety of methods of calculation. The particular merits of each method are often connected with particular applications because their originators have had such applications in mind. However, there is no clear correspondence between methods and applications. The newcomer to electromagnetic problems finds himself presented with many different methods all with enthusiastic supporters, and few critical reviews exist which compare and contrast the methods in such a way as to offer guidance as to which of them is best fitted to solve a particular problem. It is highly to be desired that such a critical dictionary of methods should be produced but this task is quite beyond the competence of the present author. In a way this book adds to the confusion by drawing attention to yet another method, which we have called the energy method. In defence we can argue that the concept of energy acts as a unifying principle, so that the method of this book enables us to see some pattern amongst the various methods. It is for this reason that this postscript to the book attempts to make a short survey of other methods. However, before setting out on the survey we must stress that the vastness of the range of application multiplied by the number of methods which already exist makes a definitive comparison almost impossible. The best advice which can be given is that a newcomer should acquire experience and skill with one or two methods and should not dissipate his early efforts amongst many different methods, however enthusiastically these are being advocated by other workers. On the other hand when experience has been acquired it is worth seeking to remedy the failings of a particular method by reference to those methods which are claimed not to have these failings.

7.2. Analytical methods

Let us begin this comparative study by assessing the role of analytical methods. Such methods are often contrasted with numerical methods and it is conventional wisdom to speak of analytical methods as belonging to the pre-computer age. This is, however, a mistaken reading of history. Even a cursory glance at Maxwell's treatise and other writings of his times shows that numerical methods were in use long before the advent of electronic computers. This is not to deny that the scope of such methods has enormously increased. However, it is unhelpful to contrast analytical with numerical methods as if one had to choose either the one or the other. It is better to regard both methods as means of providing numerical information. Electromagnetic problems are concerned with effects distributed in space and time. Analytical methods exploit the property of symmetry in such a way as to be able to store the information in a particularly compact and general form. It is, for instance, helpful to regard the circular functions such as $\sin x$ as a set of vectors with convergent sum. The numerical value of $\sin x$ is always an approximation because it is calculated by using a finite number of terms. Thus in use analytical expressions have no exact solution any more than numerical expressions, but they are generally, although not invariably, simple to compute. Besides this economy in the computation process analytical solutions often allow a separation of the dimensional variables and hence a reduction in the labour of numerical computation. For instance it is possible to reduce the three-dimensional magnetic field of a cylindrical structure, such as a rotating machine, to two dimensions by postulating a sine or cosine variation with angle and then solving for the (R, z) distribution. Such a method is a hybrid between a completely analytical solution and a completely numerical one.

Another method which successfully exploits symmetry is the method of conformal transformation. Before the advent of electronic computers this method was restricted by the small number of analytical functions which satisfied the governing differential equation. It is now possible to retain the simplicity of the method but to increase its scope by numerical integration. A further step forward is available through numerical iteration procedures which greatly enlarge the scope of the principle of superposition. Such procedures can build up complete solutions from standard parts which need some adjustment in boundary conditions before they are compatible. Of course such problems could be solved by purely numerical schemes, but experience teaches that it is far quicker to use simple component solutions derived analytically. In this view, therefore, analytical solutions should not be discarded, but should be used as stepping stones in a numerical computation. Such hybrid methods have received less attention than they deserve. Conversely it is sometimes easier to construct analytical functions such as complex Bessel functions by using a purely numerical scheme which is not based on any particular series expansion.

In problems which have a high degree of symmetry analytical methods are the obvious choice. Not only do they present what might be termed a library of sub-routines for computation which shorten the computation effort, but they also provide explicit formulae for mathematical processes such as integration and differentiation which therefore do not have to be undertaken numerically. All this makes an analytical solution of a field problem a 'complete' solution from which particular forms such as energy functionals can be easily obtained. The mathematical features of such a method are so clear that sometimes there is a temptation to forget the fact that all numerical results are approximations and that all physical phenomena are subject both to inherent inaccuracies and also to inaccuracies imposed by the processes of measurement. Thus while analytical methods should be used wherever possible, they should be regarded as a convenient means of computation rather than as a technique which enables us to dispense with computation altogether.

7.3. Relaxation methods

Engineering systems are never completely symmetrical because it is never possible to manufacture apparatus without some tolerances. Moreover, there are many degrees of freedom in the internal interactions which affect the distribution of energy in space and time. Often there are no analytical expressions which fit a particular structure. Recourse may sometimes be had to the use of infinite series of analytical terms and in this context the power of Fourier series is remarkable. It may also be possible to use perturbation methods in which the actual shape is represented by an analytical function subject to small alterations. The convergence of such solutions depends on the closeness of the real shape to the ideal theoretical expressions.

However, most problems require methods of solution which are not limited to geometrical shapes of such symmetry. Here it is important to notice that the physical processes which form the object of the investigation are generally fairly independent of the geometry. Thus a solution of a waveguide in rectangular co-ordinates presupposes sharp corners, but the general process of wave propagation is by no means limited to guides having a rectangular cross-section. The physical processes may be more easily understood when they occur in symmetrical structures but they are not determined by such structures. We therefore require general numerical methods independent of symmetry.

This line of reasoning leads to us to enquire by what means the physical statement of, say, Laplace's equation can be disentangled from the geometrical statement. Readers of this book will not be surprised when we assert that the answer is to choose some principle of invariance such as energy or action and to investigate the conditions of equilibrium by means of a variational principle. Laplace's equation is then seen to be a statement about minimum potential energy. It is Lord Kelvin's great contribution to electromagnetism that he saw this so clearly and thus influenced Maxwell in the development of electromagnetism as a branch of mechanics. The solution which Maxwell sought was by no means the solution of certain partial differential equations governing the phenomena. Rather he investigated the equilibrium of certain systems and found that the equilibrium conditions could be concisely stated by his differential equations.

This approach led Maxwell immediately to numerical methods, which depend on a guessed distribution of the energy which is then improved by a consideration of the constraints. An example which makes use of duality is Maxwell's calculation of capacitance using guessed equipotentials and guessed flux lines and then adjusting the angles between them.

This type of numerical calculation was fashioned into a design tool by Southwell, who started with the calculation of stresses in braced frameworks. Instead of calculating the displacements directly, he guessed the displacements and calculated the resultant forces. He then relaxed the displacements of the structure until the resultant forces matched the applied forces. This method was called the relaxation method and was used by Southwell long before the advent of electronic computers. He applied the procedure to Laplacian and Poissonian potential problems and paid particular attention to conformal transformations. Southwell's approach was strongly influenced by the notion of equilibrium not only in the development but in the application of the method. Thus the numerical relaxation process was undertaken by relaxing first those displacements which had produced the largest deviations from equilibrium. In this way the relaxation process could be made sufficiently convergent to allow fairly complicated systems to be solved by hand computation.

7.4. Finite-difference methods

In the solution of frameworks the displacements and forces are required at the nodes and Southwell applied the idea of nodes to the solution of

Laplacian problems. He overlaid the region in which the potential was to be calculated with a 'relaxation net' which converted the differential equation into a finite-difference equation. He argued that any physical system can be represented by a finite number of co-ordinates. This of course is the basis of Lagrangian mechanics which allows an arbitrary choice of co-ordinates, but Southwell confined his attention to small meshes in his relaxation nets so that these nets should approximate to a continuous sheet. He thus laid the basis of what are now called finite-difference methods. He also showed that many different shapes of mesh were possible and described rectangular, triangular, and hexagonal finite-difference meshes.

The advent of electronic computers made an enormous difference to Southwell's method by allowing the relaxation processes to be carried out automatically. There is now a large literature of finite-difference calculations for all the equations arising in mathematical physics. Naturally much of this work has concentrated attention on mathematical and computational features such as convergence, storage requirements, and error theory. Mathematically the object of the computation is the solution of an equation. Thus the Laplacian problem is 'solved' when the potential has been found throughout the region of interest. Such solutions contain an immense amount of numerical information and from this information it is possible to find answers to many of the questions arising in an engineering design. The regularity of the mesh size makes the computer program relatively easy to construct. Moreover, there is now available much experience in choosing the best form of the finite-difference equation for particular applications. Various schemes are available for improving relaxation speed and the method is firmly established. The only objection to it is that it attempts too much by filling all space with its net of small meshes. For many applications far fewer degrees of freedom in the co-ordinates would suffice. In this lies the advantage of the Lagrangian energy method which enables the analyst to restrict the information at the outset to those items which he requires. In that context Laplace's equation does not need to be solved but arises as an equation giving general information about the energy distribution at equilibrium.

7.5. Finite-element methods

The regularity of the finite-difference mesh size is a great advantage in programming. However, as we have already noticed this regularity may produce too much information. It might be a good idea to have a large mesh size in some parts of a system and a smaller mesh size where accurate local information is essential. Finite-difference methods can be adapted for this purpose, but the computation is complicated by the need to join the regions of different mesh size.

   The possibility of a completely free topology for the mesh is introduced by
the finite-element method which is based on the energy distribution rather
than on the differential equation describing the equilibrium condition. The
energy distribution is of course expressed in terms of a functional and the
method fits into the Lagrangian framework which we have described in this
book. The mesh is generally a triangular one and the trial functions are
generally linear in the space co-ordinates. Because the shape of the triangles
is arbitrary the method is very flexible and can be easily adapted to different
geometrical boundaries. At the same time the meshes can be made small in
those regions where the field is changing rapidly in space and large in
regions of nearly uniform field. Hence not only the energy distribution but
the field distribution itself can be obtained. The disadvantage of the method
of computation lies in the irregularity of the meshes which requires additional effort in preparing the computer program. The choice between
finite-difference and finite-element methods in field calculations is finely
balanced, depending on experience, skill in programming, computer soft-
ware availability and storage capacity, and of course the problem which is
to be investigated. To anyone skilled in the art these remarks will appear to
be painfully obvious. What is, however, less obvious is the different starting
point of the two methods. Protagonists of the finite-element method often
regard it as a method which obtains the field by discretization in a similar
but more economical manner compared with the finite-difference method.
This seems to us to be a very restricted view. To us the field is physically the
interaction of a system with a point source. It is our contention that a
knowledge of the field is required only if the action on small particles is
under discussion. In general many engineering problems are concerned with
larger pieces of material and the degrees of freedom of such systems are
reasonably small. In such systems a Newtonian particle approach is mis-
guided and a Lagrangian approach is the better choice. This means that the
field as such is not required, but instead we must find an energy functional
as is done in the finite-element method. As long as that method is regarded
chiefly as an alternative to the finite-difference method its true nature is
misunderstood. Perhaps for this reason little attention has been paid to dual
bounds and other approximation techniques which are available to the
finite-element method but not to the finite-difference method or to any other
technique which seeks a complete field solution in space and time.

7.6. Integral methods

Finite-difference methods start with the differential equations which relate
the local curvature of the field to the local source density. Finite-element
methods, as normally understood, proceed in a similar manner, although
the differential equation is first derived from a functional. Both techniques

then work outwards to the boundary, at which there are assigned boundary conditions. The computation process distributes the boundary inputs in accordance with the local constraints imposed by the differential equation.

An alternative method starts from the boundary conditions and works inwards using an integral formulation. If in an electrostatic example the boundary is specified in terms of charge density or dipole density the potential could be found at once by integration. In general, however, the boundary conditions will be in terms of potential or potential gradient and an iterative process is required. This process matches the local surface potential to the potential obtained from the distant surface values. The resultant set of equations is usually smaller in number than the set required for a volume discretization, because the number of dimensions is reduced from three to two. On the other hand these equations have a full matrix, whereas the finite-difference and finite-element schemes give an impedance matrix which has terms only along a diagonal band and can therefore be manipulated more easily.

Various methods are in use which employ integral methods. Amongst these the boundary-element method divides the boundary into finite elements.

The uniqueness theorem for static fields provides the possibility of writing the integral equations either in single- or double-layered sources. The single layer gives a gradient or Neumann condition and the double layer corresponds to a potential step or Dirichlet condition. The latter choice is advantageous because the field of a double layer decreases more rapidly with distance. The accuracy of the integral method is limited only by the accuracy with which the boundary conditions are inserted in the equations. Subject to this limitation the fields in the volume of interest are true solutions of the field equations. This is not so for the finite-difference and finite-element methods because in those methods the accuracy is limited also by the mesh size in the volume. A disadvantage of integral methods as against finite-difference methods is the requirement for a more complicated computer input specification.

7.7. Moment methods

An extremely useful general approach to methods of calculation is given by R. F. Harrington in his book *Field computation by moment methods.* Harrington's approach is mathematical. He addresses himself to the general problem of solving the operator equation

$$L(f) = g \qquad (7.1)$$

where L is an operator, f an unknown field distribution with known boundary conditions, and g a known source distribution. The method of moments is described as follows. The field f is described in the domain of L by a series of basis functions f_n such that

$$f = \sum_n \alpha_n f_n \tag{7.2}$$

where the α_n are constants. Another set of functions called the weighting functions w_m is defined in the range of L so as to form a suitable inner product with the source distribution g. This inner product results in the statement

$$\sum_n \alpha_n \langle w_m, L f_n \rangle = \langle w_m, g \rangle. \tag{7.3}$$

This set of equations can be written in matrix form. The coefficients α_n are found by matrix inversion and hence the field f is determined. The accuracy of the solution depends on the choice of the basis and weighting functions. The special case of $w_n = f_n$ is known as Galerkin's method.

It is very instructive to examine this method in the light of our knowledge of the physical structure of electromagnetic fields.

We notice first of all that electromagnetic problems are examples of the interaction of different sources such as charges and currents. This interaction can be most conveniently described in terms of mutual energy. Mathematically such an energy distribution can be described by an inner product.

If now we consider the inner product $\langle w_m, g \rangle$ of equation (7.3) we can identify g as a source distribution which has a mutual energy with another source distribution, the effect of which is described by w_m. In our previous discussion we have repeatedly found the need to introduce adjoint quantities in order to define a stationary energy functional. Let us now introduce the adjoint system defined by

$$L^a(f^a) = g^a. \tag{7.4}$$

Then the weighting functions w can be identified with the field of the adjoint sources. Thus $w = f^a$. We thus have two systems adjoint to one another and described by the operator equations (7.1) and (7.4).

The energy functionals expressing the mutual energy of the two systems are

$$W = \langle f, g^a \rangle = \langle f^a, g \rangle. \tag{7.5}$$

Physically these energies describe the process of inserting the probe g^a into the field of g, or reversing the roles and inserting the sources g into the field of g^a. In each case the action of one of the sources is expressed in terms of its field and this field is related to the source through the operator equation.

We can express this relationship by introducing it as a subsidiary condition through the method of Lagrange's multiplier. Thus

$$W = \langle f, g^a \rangle + \langle \lambda, g - L(f) \rangle. \tag{7.6}$$

Consideration of this expression shows that the multiplier λ is the adjoint field f^a acting on g. Hence

$$W = \langle f, g^a \rangle + \langle f^a, g - L(f) \rangle. \tag{7.7}$$

For stationary energy

$$\delta W = 0 \tag{7.8}$$

$$\langle \delta f, g^a \rangle - \langle f^a, L(\delta f) \rangle = 0 \tag{7.9}$$

whence

$$\langle \delta f, g^a \rangle - \langle L^a f^a . \delta f \rangle = 0 \tag{7.10}$$

and

$$L^a f^a - g^a = 0. \tag{7.11}$$

Thus the two operator equations give the necessary and sufficient conditions for a stationary mutual energy between the sources g and g^a. We note in passing that W can be expressed in a variety of forms. Harrington uses the expression

$$W = \frac{\langle f, g^a \rangle \langle f^a, g \rangle}{\langle Lf, f^a \rangle}. \tag{7.12}$$

Physically the identification of one set of sources with g and another with g^a is arbitrary.

This method is extremely powerful and shows at a glance why we can use arbitrary basis functions f and weighting functions w in the method of moments. Clearly the basis functions must be chosen to give an economical and accurate description of the field of the sources g, while the weighting functions must similarly match the adjoint sources g^a.

If the sources g describe the system under investigation, g^a describes the probe by which the original system is to be explored. One possibility is to use the sources themselves as a probe. For a self-adjoint operator such as the Laplacian operator ∇^2 this means that the two operator equations are identical. Thus

$$g = g^a, \qquad f = f^a, \qquad L = L^a. \tag{7.13}$$

This is Galerkin's method and the energy is the mutual energy of the system under consideration. This of course is the energy which we have been seeking to calculate throughout this book. It is also the energy which forms the starting point of the finite-element method.

By its definition this system energy does not give the distribution of the energy inside the system. The finite-element method overcomes this difficulty by dividing the system into a set of sub-systems. The moment method can do so by choosing probes of smaller size than the system.

The ultimate in accuracy is a probe of zero dimensions in space and time. Such a probe is mathematically described by a Dirac impulse function and its field is a Green's function. If this sharp probe is of unit strength, the energy becomes

$$W = \langle f, g^a \rangle = \langle f, 1 \rangle = f. \qquad (7.14)$$

Thus in ascribing a point value to the field f we imply that the system has been probed with a point source. This is the underlying principle behind any method of field calculation which seeks to determine a field distribution. It is, for example, the basis of the finite-difference method.

It is helpful to keep the idea of a sharp probe in mind when dealing with physical problems in electromagnetism because it focuses the mind on the relationship between calculation and measurement. It is seldom sensible to calculate to an order of accuracy which is unattainable by measurement. Nor is it sensible to calculate to an order of accuracy which has no physical counterpart because of manufacturing tolerances or because physical parameters such as permeability and permittivity are themselves averages which have an unknown point distribution.

Any scheme of calculation which is unnecessarily accurate cannot be checked physically and moreover involves additional expense and complication both in the acquisition and in the handling of the data. The moment method, based as it is on the notion of equilibrium, suggests that the probing source should be chosen in such a manner that it gives the required information and no more. Thus a mutual energy between a set of large coils does not require to be calculated by the use of a tiny coil which explores the entire region. It is only where local effects are very important that we need local information. Examples of such problems are electric breakdown problems where the exact geometry is of great significance.

7.8. Dual energy methods

The discussion of the method of moments shows that there is an enormous choice available for calculation schemes of electromagnetic energy. It is based on the idea of using an adjoint probing system. The method of moments can be used equally well with integral and with differential operator equations. However, as generally understood it leaves out one enormously useful property of electromagnetic systems. It leaves out of account that in such systems we have the possibility of describing the energy in two different ways. The theme of this book has been that there are kinetic energy and potential energy, currents and dipoles, curl sources and divergence sources. The operators which in the previous section we have denoted by L and L^a are themselves products of adjoint operators. For example the Laplacian operator

$$-\nabla^2 = -\mathbf{\nabla} \cdot \mathbf{\nabla} \tag{7.15}$$

where $-\nabla$ and $\nabla \cdot$ are adjoint.

This enables us to split the operator equation and to use both a convex and a concave functional for the energy. By this means we obtain upper and lower bounds for the energy and obtain confidence limits in an extremely economical manner. Whereas the method of moments generally operates with two parameters, the source and the field, the dual energy method operates with four parameters: co-ordinates, velocities, momenta, and forces.

All electromagnetic calculation is based on the notions of energy and equilibrium. However, the method which we have described in this book uses the insights of Lagrangian mechanics and distinguishes two types of equilibrium by examining not only the first but also the second variation of the energy. Since it has two principles to work on, it is not surprising that it can greatly reduce the computational labour.

Further reading

A compendium of analytical solutions based on the method of the separation of variables is given by P. Moon and D. E. Spencer, *Field theory for engineers*, Van Nostrand, Princeton, N.J. (1961).

A pre-computer-age text but well worth looking at is R. V. Southwell, *Relaxation methods in engineering science*, Clarendon Press, Oxford, Vol. 1 (1940) and Vol. 2 (1946).

An excellent introduction to finite-difference methods is G. D. Smith, *Numerical solution of partial differential equations*, Clarendon Press, Oxford (1965).

The standard text on finite-element methods is O. C. Zienkiewicz, *The finite element method in engineering science*, McGraw-Hill, New York (1971; 3rd edn., 1978).

A lucid account illustrated by many examples is given by R. F. Harrington, *Field computation by moment methods*, Macmillan, New York, 1968.

Index

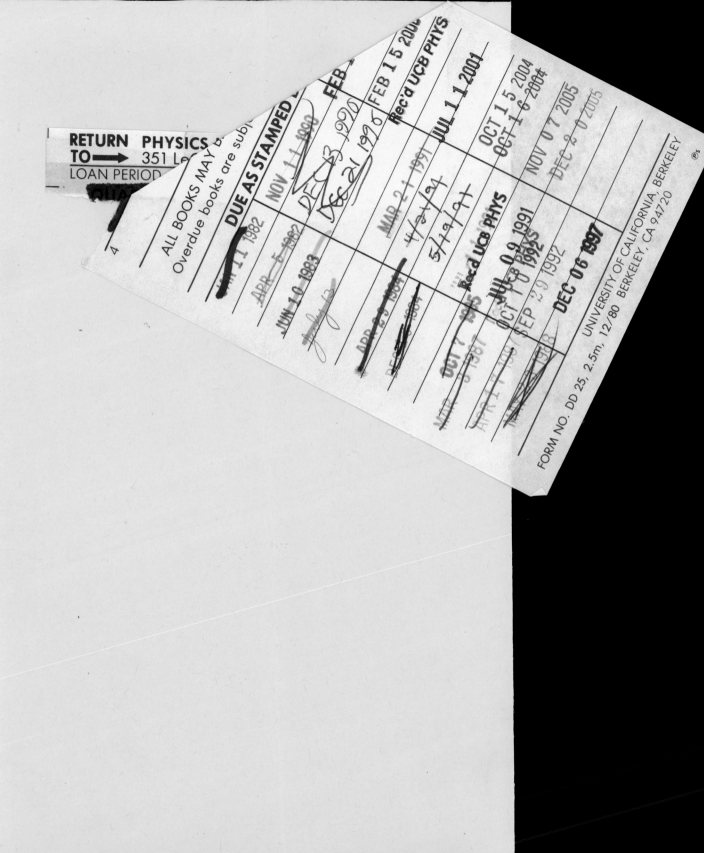